The Star of Bethlehem

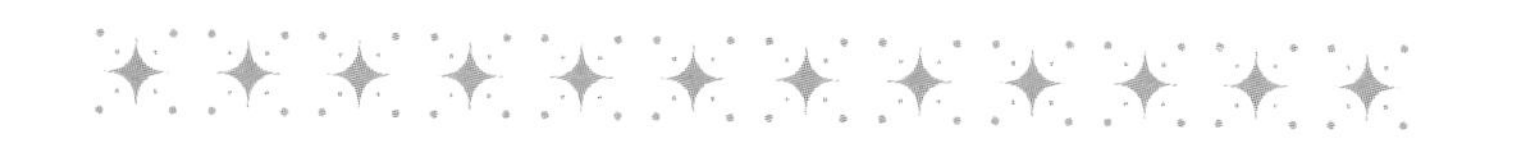

The Star of Bethlehem

AN ASTRONOMER'S VIEW

Mark Kidger

PRINCETON UNIVERSITY
PRESS

Published by Princeton University Press, 41 William Street,
Princeton, New Jersey 08540
In the United Kingdom: Princeton University Press,
Chichester, West Sussex

Library of Congress Cataloging-in-Publication Data
Kidger, Mark R., 1960–
The star of Bethlehem : an astronomer's view / Mark Kidger.
p. cm.
Includes bibliographical references and index.
ISBN 0-691-05823-7 (cloth : alk. paper)
1. Star of Bethlehem. 2. Astronomy in the Bible.
3. Astronomy, Ancient. I. Title.
QB805.K525 1999
523.8′446—dc21 99-29635

This book has been composed in Centaur

The paper used in this publication meets
the minimum requirements of
ANSI/NISO Z39.48-1992 (R1997)
(*Permanence of Paper*)

http://pup.princeton.edu

Printed in the United States of America

3 5 7 9 10 8 6 4 2

 CONTENTS

PREFACE
vii

Chapter 1. Matthew's Star
3

Chapter 2. A Star over Bethlehem?
20

Chapter 3. The First Christmas
39

Chapter 4. Halley's Comet and Other Red Herrings
73

Chapter 5. Shooting Stars and Fiery Rains
110

Chapter 6. Supernova Bethlehem?
136

Chapter 7. We Three Kings?
166

Chapter 8. Triple Conjunctions: A Key to Unlocking the Mystery?
198

Chapter 9. Is the Answer Written in Chinese?
219

Chapter 10. What Was the Star of Bethlehem?
247

Epilogue. Which Star Is the Star?
267

Appendix. The Heavens above Bethlehem
277

NOTES
289

BIBLIOGRAPHY
295

INDEX
301

PREFACE

The Star of Bethlehem has been a mystery for many centuries. At Christmastime it is usually discussed around the world by scientists and nonscientists alike, all of whom are fascinated by its meaning and its history. Millions of people are reminded of it annually, for the Star of Bethlehem is one of the most popular illustrations on the front of Christmas cards.

But what was this mysterious apparition in the heavens? Will we ever know? After all, it is mentioned in only one of the four Gospels, and then only briefly. Even so, tens of thousands of printed pages have been devoted to it. There are opinions for all tastes: some people say that the Star never existed but was added to the Bible to give significance to the birth of Jesus; others feel it was a simple astronomical event; and many Christians feel it was a miracle, a sign placed in the heavens by God to indicate the divine nature of the infant Jesus. There are dozens of theories and hundreds of opinions—and each year these numbers increase.

Scientists, too, have been captivated by this mystery. Assuming that the Star was a real astronomical event, they have sought to explain it in scientific terms. One of the first to do so was Johannes Kepler, the great astronomer who lived and worked at the end of the sixteenth century and beginning of the seventeenth. He wrote about the Star of Bethlehem after just having seen a spectacular conjunction of planets followed a short time later by a bright supernova. Kepler, however, was far from the first to ponder the nature of the Star. Certainly, there is good circumstantial evidence that its nature was debated by theologians at the turn of the fourteeth century. But there are even documents written a thousand years earlier that mention the Star of Bethlehem, and some of these discuss, albeit superficially, what the Star could have been.

If the biblical account is taken literally, then no scientific explanation is necessary or even possible; it could only be a miraculous event. But behind the few verses of the Bible that mention the Star, there is a fascinating story with constant twists and turns. We must wrestle with records written long after the events they outline, and with accounts written by people who could not possibly have seen the things they describe and discuss. The accounts of the Nativity were recounted orally, time and time again, passing through the decades until they were finally written down toward the end of the first century. Even though there was a long-standing oral tradition, it is possible that the story of the Star changed—perhaps subtly, perhaps substantially—in the retelling. Alas, we shall probably never know how much, or in what ways. Certainly the content of the Gospels shows that their writers adapted their style according to the people for whom they were writing.

There are many people who feel that the Star of Bethlehem is an ordinary mystery, to be tackled the same way that scientists explore other mysteries of nature: investigate, reason, hypothesize, and conclude. Others feel that the mere attempt to explain the Star of Bethlehem in this manner is to reduce its importance and is therefore virtually blasphemous. A third group would prefer to doubt that the Star even existed, so in their view any attempt to discuss it is a waste of time and energy.

The Star of Bethlehem is perhaps the greatest of all astronomical mysteries; it is certainly the most enduring and one of the hardest to solve. Whether one thinks it was a miracle closed to scientific analysis, a simple chimera, or an astronomical event to be explained like all other natural phenomena, it remains a fascinating story. Probably we will never know, and *can* never know, for certain what the Star of Bethlehem really was: two thousand years later the records are so old, incomplete, and circumstantial that we may never be able to read them correctly or draw the right conclusions from them.

In this book I will try to put the scientific evidence in context and to separate the possible from the impossible. Sometimes we will find that a particular theory is based on an error or a misunderstanding, other times that a theory is possible but implausible. This book is an astronomer's search because I am an astronomer first and foremost, although the sheer mystery of the Star has intrigued me for many years as well. The book would not be complete, however, without a discussion of some background events: biblical history, the history of long-disappeared civilizations, and religious and astronomical practices. An astronomer may not be the best qualified person to take on such a task, particularly in areas where even the greatest experts have profound and fundamental differences, and where many of the agreed-upon facts are based on the penetrating and complicated interpretation of subtle clues. In these cases one has to rely on available sources and expert opinions, which one trusts to be reliable and authoritative.

Our exploration requires us to marshal evidence from many different time periods and from different cultures. We must analyze texts written in Aramaic, Greek, and Chinese, and examine studies of the planets and stars. In our journey we will encounter eclipses, comets, and new stars. We will chase red herrings and follow false trails, sometimes misled by a single word or even a single character. But it will still be a rewarding journey. I believe that we are as close as we will ever be to understanding what the Star of Bethlehem was—assuming, of course, that the mystery can ever be resolved using sound science, logic, and educated guesswork.

Over the years I have presented the argument of the book, heavily summarized, in numerous talks and articles. During this time it has been revised and refined to some degree, although without fundamental changes until it has reached the version presented here. Very rarely—just once so far—has someone felt offended by the conclusion that the Star of Bethlehem has a fairly simple astronomical explanation. The reason is simple: those who seek a scientific explanation will find one; those who want a miraculous explanation will be satisfied that the Star *did* exist and that many of the events described in the New Testament can be proved historically; and, finally, the full explanation requires a singular—one could say miraculous—combination of events.

Some may, not surprisingly, find the theory controversial or unacceptable for one reason or another. But where facts are hard to establish and much depends on opinions, a tolerance for different points of view is essential. Recently, quite by accident, I met up with a champion of one of the alternative explanations. We sat on a sofa and discussed the ins and outs of both theories over coffee, and finally agreed to disagree in our conclusions. If only all scientific debates were so cordial!

Whatever your opinion, whether you find the reasoning expressed here convincing or not, I hope that you will enjoy reading this greatest of all detective stories.

March 1999
La Laguna, Tenerife

The Star of Bethlehem

I

Matthew's Star

It is Palestine, sometime in the decade between A.D. 85 and 95. An old man is trying to write down what will be a synopsis of events that happened a long time ago. His work is a memoir of his experiences while he was a young tax collector in this distant outpost of the pagan Roman Empire. He writes little about himself, concentrating mainly on one who more profoundly walked these lands three generations ago. The writer's name is Matthew. He is writing about Jesus of Nazareth, the one who, when they were all so young, was cruelly crucified on the hill later to be called Calvary. Back then it was called Golgotha, or simply, "the place of skulls."

But Matthew is still far from describing the day of the crucifixion, nearly the final story in his synopsis. Nor has he even reached the point of his own modest entry into this history, recounted in Matthew 9:

> 9 As Jesus was walking along, he saw a man called Matthew sitting at the tax booth; and he said to him, "Follow me." And he got up and followed him.

He is grappling with descriptions of the time just following the birth of Jesus. It is the beginning of the whole story of

Christianity, one that we are still reading two thousand years later. Matthew's is a concise narrative, some action, much emotion, story of great deeds; but there is hardly any description of place, except for a single astronomical event—an event that does not show up in any of the three other accepted accounts of the life of Jesus that we now call the Four Gospels.

According to Matthew 2:2,

> 1 In the time of King Herod, after Jesus was born in Bethlehem of Judea, Magi from the East came to Jerusalem,
>
> 2 asking, "Where is the child who has been King of the Jews? For we have seen his star at its rising[1] and have come to pay him homage.

And Matthew 2:7–11,

> 7 Then Herod secretly called the Magi and learned from them the exact time when the star had appeared.
>
> 8 Then he sent them to Bethlehem, saying, "Go and search diligently for the child and; when you have found him, bring me word so that I may also go and pay him homage."
>
> 9 When they had heard the king, they set out; and there, ahead of them, went the star that they had seen at its rising, until it stopped over the place where the child was.
>
> 10 When they saw that the star had stopped, they were overwhelmed with joy.
>
> 11 On entering the house, they saw the child[2] with Mary his mother; and they knelt down and paid him homage. Then, opening their treasure-

chests they offered him gifts of gold, frankincense and myrrh.

Matthew's story comes not solely from an old man's memory; in fact, there is grave doubt that Matthew, the tax collector, the apostle of Jesus, even wrote the text of the Gospel attributed to him. But he may have been one of its sources; some of the rest of the story came from written accounts, some came from the very early Christian oral tradition.

What is striking, and what is not observed in the other three Gospels, is the importance given to this Star that marked the birth of Jesus. In fact, the Star is not even mentioned in the Gospels of Mark, Luke, or John. But Matthew's account of the Star—or that of the writers of his account—persists for several lines, literally drawing for readers the arc of the Star's rising and setting above "the place where the child was."

This is at once a strong and troubling image. Why is there no mention of the Star in the other Gospels if it was such an important part of the Nativity? A partial answer: the Gospels were written by different people and intended for different audiences. Furthermore, besides the Four Gospels that we know so well—Matthew, Mark, Luke, and John—there were also the so-called Apocryphal Gospels, which were not included in the definitive compilation of the New Testament. There are probably others, still undiscovered, and there was an oral tradition as well. In effect, there are many stories of Jesus shared among several early Christian sects.

To interpret Matthew's account of the Star properly, we must take a closer look at the origin of his Gospel's narratives. If we can understand what emphasis he wants to place on the events he records, we will be in a better position to determine how much significance should be given to the report of the Star.

Anyone who has read the Four Gospels will notice one thing immediately: John's is unlike the other three. It is clearly intended as a more theological, spiritual, or mystical account than as a "biography." In fact, the fourth Gospel was so different that it was one of the last books to be accepted into the New Testament canon, only after Irenaeus, the second-century bishop of Lyons and arguably the church's first great theologian, staunchly defended its authorship. In John's account, we first encounter Jesus at his baptism in the river Jordan, an event that marks the beginning of his public ministry. The Gospel of John does not even describe the infancy of Jesus—no stable, no shepherds, and no Star.

The other three Gospels have, broadly speaking, similar descriptions of the events of Jesus' life. For this reason they are called the "Synoptic Gospels," from the Greek word first coined by Plato and used in his *Laws* and the *Republic,* meaning "to bring into a general view"—the same root as the word "synopsis." The degree of interdependence among Matthew, Mark, and Luke, as well as the order in which they were written, has generated what is known as the "synoptic problem." In attempts to resolve these issues, scholars have applied techniques from literary criticism and textual analysis. The results are revealing and also suggest why only Matthew's Gospel includes any mention of the Star.

To begin their study, scholars wrote out in abbreviated form the events recorded in the Synoptic Gospels. Two things became apparent from this exercise. First, even though the same events may have been recorded in the three Gospels, they did not necessarily occur in the same order. For example, Mark and Matthew may agree on the order of a sequence of events, and Luke disagrees; or Mark and Luke may agree, but Matthew disagrees. But, rather like the dog that didn't bark in the Sherlock Holmes story, it never happens that Matthew and Luke agree and Mark disagrees. Second, looking closely at the wordings of particular episodes in the Synoptic Gospels, scholars found that the same thing holds true here as well: Matthew and Luke tend to agree on wordings, but Mark disagrees.

These two findings strongly suggest that both Matthew and Luke had knowledge of Mark's account and used it when writing their own. In other words, the Gospel according to Mark was the first to be written down. Other, indirect evidence supports this view. For example, Mark's account contains some phrases that were likely to have been thought of as "offensive" at the time, and they are toned down or omitted in the other two Synoptic Gospels. It also contains some Aramaic terms and has a less polished literary style than Matthew's and Luke's accounts. But, even if it is the earliest Gospel, who wrote it, and when?

There is nothing in Mark's Gospel that identifies the author. The Greek texts contain many Latin loan words, which makes it plausible that the writer was living in Rome at the time. In the early second century, Papias, who was bishop of Hieropolis in Greece between about A.D. 130 and 140, is reported to have said:

> Mark, Peter's interpreter, wrote down what the Lord had said and done—as far as he remembered it—accurately, but not in order. For he had neither heard

> the Lord nor followed him, but later, as I said, he was a follower of Peter, who gave such instructions as circumstances required, and not an orderly account of the Lord's words. Hence Mark was not at fault in writing some things simply as he remembered them. For his one care was to omit nothing that he had heard, and to speak truthfully thereon.

This passage, from Eusebius's *History of the Early Church*, identifies the author as John Mark, to whom Peter addressed his first Epistle. Irenaeus, who defended the authenticity of John's Gospel, writes in *Against Heresies* that "after the departure of Peter and Paul, Mark—Peter's disciple and interpreter—delivered to us in writing what Peter had preached." But, although the early church may have assumed that Mark was the author, modern scholars are not convinced—even if Peter does play a more important role in Mark's Gospel than in the others. Traditionally, however, it is thought that Mark's Gospel was written in Rome, shortly after the death of Peter, and hence between about A.D. 64 and 67. Some scholars think that the thirteenth chapter of Mark does not presuppose that the Temple in Jerusalem had been destroyed. It is known from external, nonbiblical sources that the Romans seized Jerusalem and destroyed the Temple in A.D. 70, which, if the above interpretation of Mark, chapter 13, is correct, means that it could not have been written later than this date. However, other scholars think chapter 13 was written down after the destruction of the Temple, and so they place the date of writing shortly after A.D. 70.

Matthew and Luke pose further problems. About 80 percent of Mark's account is reproduced by Matthew, and about 65 percent by Luke. But even if we extract Mark's material from Matthew's and Luke's accounts, their Gospels still have a lot in common. Much of their material is focused exclusively on the sayings and teachings of Jesus, leading scholars to suggest that both Matthew and Luke relied upon another narrative besides Mark's—either a written document, now lost, or simply a different oral tradition. This additional "sayings" narrative is known as "Q," from the German word *Quelle,* meaning "source." Interestingly, the tentative "table of contents" that experts have constructed for Q does not usually include the Nativity passages, even though Mark's and Luke's Gospels are the only ones to discuss the birth and infancy of Jesus.

The Gospel of Luke and the New Testament book Acts of the Apostles are traditionally attributed to one of the apostle Paul's co-workers, described in Colossians as a physician. Again, modern analyses of the texts cast doubt on this assumption, in spite of a nineteenth-century claim that Luke-Acts do use specialized first-century medical language. For example, the author-physician conflicts with Paul on several points, and the theological statements attributed to Paul in Acts differ from those expressed in his letters. This discrepancy does not seem to square with an author who is a close companion of Paul! Furthermore, it is known that Luke and Acts were written after the destruction of the Temple by the Romans, one of the major events, along with the Zealots' "last stand" at Masada, of the Jewish wars described in the book by the Jewish historian Josephus (ca. A.D. 37–100). Finally, the author has a sense of "church" and of "theology," which did not evolve among the Christians until a little later. Bible historians therefore think Luke was written around the year A.D. 85, between ten and twenty years after Mark.

Then, of course, we must consider the Gospel of Matthew itself. Again it is Bishop Papias who first comments—some fifty to eighty years later—on its authorship: "Matthew wrote the Oracles of the Lord in the Hebrew Language; but everyone interpreted them as best he could." Irenaeus, probably borrowing from Papias, says that "Matthew published his Gospel among the Hebrews in their own tongue." Both Papias and Irenaeus, when they use the word "Hebrew," almost certainly mean Aramaic, a language that was spoken in Babylonia and superseded Hebrew in Palestine, at the time the Jews returned from their captivity in Babylon. Modern scholars argue against Irenaeus and Papias. Matthew's Gospel is clearly not a set of sayings (oracles), and the Greek texts seem far too good as literature to have been translations. In fact, Papias's comments seem to apply better to Q than to Matthew.

The text of Matthew also seems to assume that the destruction of the temple has already taken place. Furthermore, the Gospel appears to be cited by Ignatius of Antioch in his letters to the Smyrnans and to Polycarp, written in the early part of the second century. Consequently, we can conclude that Matthew's Gospel was probably written down in about A.D. 90.

Even though the Gospels were probably written at about this time, the earliest surviving copies of the Bible come from much later. The two oldest complete Bibles date from the fourth century A.D.: the Codex Vaticanus (held in the Vatican) and the Codex Sinaitacus (in the British Museum). Some much older fragments of the Gospel are extant, however. One of the oldest New Testament fragments is from Matthew's Gospel and dates from around A.D. 200, more than a century after the original Gospel was presumably written; it is held by Magdalen College at Oxford University. Thus, even the very oldest Gospel texts we have are probably still copies of copies.

If modern biblical scholars are correct, it is not surprising that the Star is mentioned only in Matthew's Gospel. After the death of Jesus, his followers went preaching and teaching throughout the Roman Empire. They would have described the things that Jesus did, and recounted—as best they could remember—the things he said. With the death of the first apostles and with the destruction of the spiritual and cultural center of the Jewish people (of which the early Christian community regarded itself a member), the time was right to create a written record of events, deeds, and words: the Gospels were born. In many respects, the Gospels are not "original" works, even though they represent a unique literary style. They were the selected, edited, and abridged versions of a number of long-standing oral traditions.

As NOTED, the two Gospels that describe the Nativity are those of Luke and Matthew, though their accounts are somewhat different. Luke's Gospel is the closest to what we would call historical; even then, it is written as a work of faith. For Luke, the salvation of which Jesus spoke was for everyone, no matter what one's social status. His Gospel places great emphasis on the role of women in the life of Jesus, and perhaps for this reason he provides a detailed account of the Nativity, in which Mary plays a prominent role.

Some of the most famous episodes in the story of the

Nativity are mentioned by Luke, and *only* by Luke. Unlike the rather more summarized account given by Matthew, he often gives a wealth of detail, including descriptions of the census decreed by Caesar Augustus, the journey of Mary and Joseph to Bethlehem (both mentioned in Luke 2:4), and brief details of the Nativity itself. Luke also speaks in detail of the apparition of the angel to the shepherds and of their subsequent homage to the baby Jesus. At no point, however, does Luke mention the Magi or the Star of Bethlehem.

The lack of common elements in descriptions of the Nativity as given by Matthew and by Luke is somewhat worrying if we are really dealing with factual accounts. For example, Matthew mentions the visit of the Magi while Luke mentions only the visit of the shepherds. Common explanations claim that these two events, unlike the traditional scene depicted on Christmas cards and in Nativity plays, were far from simultaneous. Some experts suggest that the visit of the shepherds came immediately after the birth of Jesus, while the visit of the Magi came weeks, months, or maybe even more than a year later. The debate centers around the Greek word used by Herod when he discusses Jesus with the Magi: is it best translated as "infant" or "toddler"? "Toddler" would suggest that Luke deals only with the events immediately surrounding the time of the Nativity, while Matthew encompasses a wider expanse of time, but in less detail. This explanation, however, does not account for Luke's omission of the Star. One must argue that it is only later, when the Magi arrived in Jerusalem to search for the baby Jesus, that the Star assumes importance in the Nativity story. If this were true, it would give strong support to those who argue that the Star of Bethlehem was not a brilliant, widely observed, or, at least, a widely known phenomenon.

Let's take a closer look at the character of Matthew's Gospel. Matthew appears to have been a Jewish Christian, that is, he viewed Christianity as the new religion in transition from Judaism, and not in opposition to it. Hence, Matthew's narrative is inclusive of the Jewish sacred texts, and he includes many quotations from the Torah. He suggests throughout his Gospel that several of the events in Jesus' life are the fulfillment of the words of Jewish prophets:

> . . . and so the prophecy might be fulfilled.

He adds this phrase onto several accounts in his Gospel. For example, it concludes the story of the flight of Mary, Joseph, and the child Jesus from Bethlehem into Egypt to escape the massacre of the innocents as decreed by the intensely insecure king, Herod. But for all of his attempts to draw a continuity of belief and narrative from Judaism to Christianity, Matthew, or whoever wrote his account, misses an opportunity to highlight another prophecy—the one concerning the Star. This prophecy of the soothsayer Balaam, spoken centuries before the birth of Jesus, is found in one of Balaam's oracles, and Matthew, being in command of Jewish scripture, would have known about it. It occurs in Numbers 24:17.

> I see him, but not now; I behold him, but not near—a star shall come forth out of Jacob and a sceptre shall rise out of Israel; it shall crush the borderlands of Moab and the territory of all the Shethites.

This prophecy has been interpreted as indicating that a star would appear at the birth of the Messiah, the anointed one of God. The passage forms one of the last of the oracles of Balaam, a series of seven prophecies made before his falling out with Balaq, the king of Moab, a region on the banks of the Jordan facing Jericho. Balaq had sent his emissaries to summon Balaam to Moab, on whose borders the Israelites had apparently begun to settle in large numbers. This situation was threatening to King Balaq because Moses' army had recently defeated Sihon, king of the Amorites to the west, and Og, king of Bashan to the north. Knowing that the soothsayer Balaam believed in Yahweh, the God of the Israelites, King Balaq thought he might be able to thwart an Israelite attack on his own lands if Balaam would foretell that the Israelites were cursed and that disaster would befall any further movement into Moab.

With great reluctance, Balaam agreed to travel to Moab, where he had altars built and animals sacrificed. Balaq promised riches to Balaam, too, if only he could denounce the Israelites. But in the end, all Balaam could foretell in his oracles was the destruction not only of Moab's inhabitants but also of others who lived on the borders of Canaan. In yet another historical example, a king took out his rage on the messenger of bad tidings: Balaq banished Balaam from Moab.

This particular oracle of Balaam offers the kind of continuity that Matthew applies throughout his account. Balaam foretells that "a star shall come forth out of Jacob." It was from Jacob that the House of David would descend, and from the House of David the Messiah was to come.

It is not clear when Balaam's oracles were written. The five books of the Old Testament which form the Torah

(Genesis, Exodus, Leviticus, Numbers, and Deuteronomy) were compiled in their present form as late as the fifth century B.C., but it seems that Numbers included some old writings as well as more recent ones. It has been suggested that the writer who added the final details to this book did so in the sixth century B.C. However, Deuteronomy, the last of the five books (and the one immediately following Numbers), may date back to the eighth century B.C.

Jewish tradition sees the oracle of Balaam in a different way. The Jewish translation of the passage given above is substantially the same, although it has some significant differences:

> I see him, but not now; I behold him, but not nigh: There shall step forth a star out of Jacob, And a scepter shall rise out of Israel, And shall smite through the corners of Moab, And break down all the sons of Seth.

Joseph Hertz, the late chief rabbi of the British Empire, interprets this passage as probably applying to King David, the first monarch to conquer King Moab, rather than to the Messiah. Later, this same passage was thought to refer to Bar Cozeba, the Jewish resistance leader during the last Jewish war of independence against Emperor Hadrian of Rome. Bar Cozeba's name was even changed to Bar Cocheba, which means "the Son of a Star," after the war. The scepter is assumed to represent someone who holds a scepter, that is, a ruler or monarch.

The absence of the phrase "that the prophecy might be fulfilled" in the verses of Matthew, chapter 2, that deal with the Star may be significant. If Matthew did not regard the oracle of Balaam as prophesying that a star would mark the birth of the Messiah, then the inclusion of the Star in his Gospel narrative becomes important: it would have been included for other than scriptural reasons. As we shall see, the reasons may have been astrological. But he may have included the phrase also because something truly unusual had genuinely appeared in the heavens around the time of Jesus' birth, and the event had been incorporated into the oral histories. Later it would form part of Matthew's Gospel.

We have to bear in mind, however, that some writers have suggested that the inclusion of the reference to the Star of Bethlehem in the New Testament was an editorial decision to give added importance to the Nativity, by ensuring that what was written in the oracle of Balaam should be seen to have come to pass. This view needs to be taken seriously, especially as the other Gospels do not mention it. However, Matthew's account of the Star is not the only one that exists—it is mentioned in at least two other ancient documents, even though they are far less known than Matthew's Gospel.

One other description of the Star of Bethlehem is far more detailed in some respects than Matthew's brief comment. It occurs in an Apocryphal Gospel, the Protoevangelium of James. Although the New Testament includes only four accounts of the life and works of Jesus, other accounts, dating from the same period, were also written down. These, too, were said to have been written by disciples of Jesus or their close followers, and include Gospels attributed to James, Thomas, and Nicodemus. When the church defined the New Testament canon, however, these Gospels were not included in the list, which had formed part of a letter of A.D.

367 by Athanasius, a bishop of Alexandria, of "books that are canonized handed down to us and believed to be divine."

The Apocryphal Gospel of James mentions the star when it reports Herod's questioning of the Magi:

> And he questioned the Magi and said to them: "What sign did you see concerning the new born King?" And the Magi said: "We saw how an indescribably great star shone among these stars and dimmed them, so they no longer shone, and so we knew that a King was born for Israel. And we have come to worship him," and Herod said: "Go and seek and when you have found him, tell me, that I may also come to worship him." And the Magi went forth. And behold, the star which they had seen in the east went before them, until they came to the cave. And it stood over the head of the child.

This account is similar to Matthew's in various ways, but gives more details of the Star and of the reasoning employed by the Magi when making their journey. It is possible, given the above remarks about the origin of the canonical Gospels, that Matthew and James were simply copying each other or using a common source. In that case, this is not an independent account of the appearance of the Star. However, the extra details included in this account may be important—in particular, because the Protoevangelium implies that the Star was extremely bright.

A slightly later record is found in the Epistle XIX of Ignatius, an early bishop of Antioch, written to the Ephesians at the start of the second century A.D., about thirty years after Matthew's Gospel:

> Its light was unspeakable and its novelty caused wonder.

This comment is brief and to the point, although we have no idea what Ignatius's source of information might have been. As with the Apocrypha, but unlike Matthew, it suggests the Star was an extremely prominent object in the sky. The clear implication of this brief passage is that the Star had appeared suddenly and unexpectedly. Not many types of astronomical objects can do this, and, as we shall see, bright objects that appear so suddenly rule out many of the possible scientific explanations of the Star.

We thus have two short biblical references to the Star, both in Matthew, and two other references in other documents written around the same time as his Gospel: the Protoevangelium of James and the letter of Ignatius to the Ephesians. The two biblical accounts say little about the brightness of the Star, while both nonbiblical references say it was very bright. One of the latter accounts (i.e., the letter of Ignatius, written the longest time after the Star) suggests its appearance in the sky had been a sudden and unexpected event. But none of the accounts says much about the nature of the Star, nor do they offer much in the way of clues.

Interestingly, at the start of the third century A.D., Origen (ca. 185–254), a prolific writer and one of the most colorful characters of the early church, wrote that the Star was "A new star unlike any of the other well-known planetary bodies . . . but partaking of the nature of those celestial bodies which appear from time to time, such as comets (etc.)." He wrote this comment around A.D. 248 in his work *Contra Celsum* and is best known through the

1953 English translation by Henry Chadwick. Origen's comments appear to be the first actual discussion of the true nature of the Star of Bethlehem, although it comes so long after the actual apparition of the Star that it is a doubtful guide to us. Though it is impossible to determine the sources on which Origen based his analysis, they appear to rule out completely some explanations that have been popular over the years, as we will see later.

The suggestion in the Protoevangelium that "an indescribably great star shone among them and dimmed them, so they no longer shone" is interesting. The British astronomer and astrohistorian David Hughes, whom we will meet often in this book, has pointed out that if this account is literally correct, the Star would have had to be exceptionally bright, at least equivalent to the moon to be able to "dim the stars." If the Star of Bethlehem really was so bright, then surely astronomical records of other ancient peoples, such as the Chinese or Koreans, would contain references to it. It would also be unusual for King Herod not to have seen it, though we can't be sure: only the Protoevangelium account states explicitly that he had not. Matthew merely states that Herod had asked diligently "what time the star had appeared."

Was the Star really brilliant? Did these early accounts use artistic license? Which of the accounts, if any, was the "correct" one? Were we even supposed to take the story of the Star literally? The Bible and the Apocryphal Gospels were never intended to be exact histories of the life of Jesus. We must remember that they are works written by the faithful for the faithful, and for those whom the writers hoped to convert.

2

A Star over Bethlehem?

When we consider what little information we have from the biblical account of the Star of Bethlehem, we can see that there are only three options available to explain the mystery:

1. The Star is a myth or a legend. It was added to the Gospel of Matthew to fulfill the Old Testament prediction by Balaam that a star would announce the birth of the Messiah. Alternatively, perhaps it was mentioned to give added significance to the birth of Jesus. In either case, the Star of Bethlehem did not actually exist.
2. The Star is a report of a genuine, normal astronomical event, modified over time by retelling or by artistic license or both, and finally incorporated into Matthew's Gospel. It is basically a true account of some phenomenon that was observed in the sky around the time of the Nativity.
3. The Star was a miracle, which is beyond the province of science to explain. In this case, the Star existed and was seen, at the time of the Nativity, but

> was not a natural phenomenon of any kind. Unlike astronomical phenomena that have been visible from Earth, and are still "up there" somewhere, the Star of the Magi has disappeared forever.

These options are well known and have been around for many years, forming the basis of the debate about the nature of the Star. Basically, they are mutually exclusive. In two of the three possibilities, the Star genuinely appeared in some way, shape, or form. Each of the three options is supported by some scholars who have studied the problem. The main arguments for each can be summarized briefly and simply as follows.

The Star Is a Myth

The biblical account was written in an age when no great king or emperor was born or died without a "celestial manifestation" to mark the event. In Shakespeare's *Julius Caesar* (act 2, scene 2, lines 30–31) we find Calpurnia, Caesar's wife, warning Caesar of the portents that she has seen in her dreams:

> When beggars die, there are no comets seen;
> The heavens blaze forth the death of princes.

Such belief in portents was much a part of Roman life. In the letters of Pliny the Younger (A.D. 61–113), we encounter frequent references to the divine status of deceased emperors. For example, in his letter to Titius Aristo (*Complete Letters,* book 5, letter 3), he refers, in one brief passage, to a whole series of recent additions to the Pantheon of Olympus: "the deified Julius Caesar and the deified Emperors Augustus and Nerva and Tiberius Caesar." Not everyone reached the extremes of naming so many deified emperors in so few words. Such references, however, were typical and became so hackneyed in Roman literature that some translators even prefer the translation "the late emperor" or "the deceased emperor" to the one given here. When all emperors effectively came to be deified upon death, the phrase ceased to have any special meaning at all. The sign that an emperor had been deified was the sighting of some type of portent immediately after his death: a comet, a strangely formed cloud, the plume from a volcano. This sign would be popularly interpreted as his soul rising to join the gods of Olympus. As has been cynically pointed out, once the precedent had been established, someone somewhere always saw something that could be interpreted as a portent of this type—that much was guaranteed!

The level of belief in portents was such that a great ruler, be he king, emperor, or simply tyrant, could only be regarded as such if his reign *had* been announced by a suitable event. Given that Matthew was writing for a public so deeply influenced by Roman values and traditions, he may have felt some obligation to follow the Roman way. Perhaps Matthew included the description of the Star only to give his account of the Messiah credibility among the pagans whom he wished to convert. Had he not done so, the story of the Nativity would have run the risk of not being taken seriously. As Jesus was the greatest ruler of all, the sign that announced his coming would have had to be especially important or spectacular.

The Star Was a Miraculous Event

The events described in the Bible are an account of the works of God. The Bible describes the Conception as a miracle, the work of God. God does not have to justify his acts and can make a new star shine in the sky if this is His way.

Science and religion have not always maintained good relations. Theologians are uncomfortable with the scientific tradition of questioning everything, because proof denies faith. There is nothing in the biblical account of the Nativity stating that the Star was a natural event rather than a work of God; as such, it is not an object to be analyzed and explained.

Various aspects of the biblical account are impossible to explain scientifically. A normal star could not appear suddenly at precisely the right moment. It could not move to guide the Magi. It would not stop over the place where Jesus was lying. The Star, according to the common interpretation of Matthew (for example, that of David Hughes), appears and disappears at different times. None of these things can be explained by a simple scientific hypothesis, nor even by a combination of natural phenomena, hence the Star was, by definition, a miraculous event.

The Star Was an Astronomical Event

It can be argued, with some justification, that a large part of the Bible was written as a historical record of events, albeit presented from an evangelical point of view. Nobody, or almost nobody, defends the position that the Bible is a straight history of events. It is undeniable, however, that a significant fraction of the Bible, both in the Old and the New Testaments, is based on true events and on people who really existed, giving a mixture of faith, vision, and history that are often difficult to separate.

Many of the events described within it have been confirmed independently. Key persons in the Nativity are known from independent sources to have lived at that time, such as King Herod. Key events such as the census decreed by Caesar Augustus are known to have happened. Given this high degree of confirmation of a large part of the biblical account by the historical record—which is unarguable—it is not unreasonable to think that the reference to the Star is also a factual account.

Experts examining the story related by Matthew and others have decanted for various astronomical events that are known to have occurred in the last few years before Christ and that could have explained the Star of Bethlehem. Phenomena we will learn about later on, such as conjunctions, novas, comets, and even occultations of the planets by the Moon, have been suggested on different occasions by different writers. Any or all of these are serious candidates to explain the Star. For these experts, the argument continues; it is just a matter of time before the correct candidate is identified.

The only significant problem with the astronomical interpretation is that the records are not eyewitness accounts. For this reason, they suffer from distortions, exaggerations, and contradictions, which, taken separately or together, may make it impossible to decipher the Star's true nature. As such, there are many astronomical events that make plausible explanations because, so far, we have been unable to discriminate adequately among them. We will begin this exciting search, and we will follow scientific rules as best we can, to explore what is certainly at times a mysterious journey.

Unfortunately, in the case of an event such as the Star of

Bethlehem that occurred so long ago—one for which we have only ancient records and no direct observational records to analyze—modern scientific methods meet a serious obstacle. There are no photographs of the Star that can be processed by computer and no telescope captured its image for storage on magnetic tape. We do not even have sketches or drawings, however rough, to ponder over. In such circumstances, the burden of proof is a heavy one. It is almost impossible to prove or disprove most of the many theories that have been proposed unless they really fly in the face of the facts.

We can explore the idea of the Star as a true astronomical event more fully if we take the Gospel accounts to be almost literally true. Doing so will permit us to find more fascinating clues, although we risk overinterpreting the data. In Matthew 2:2 we read in many of the older translations of the Bible that "we have seen his star in the east." Grammarians will immediately recognize "in the east" as a dangling prepositional phrase, for this comment can be read in at least three ways: Was it (1) the Star, (2) the Magi, or (3) both that were in the east? Recent translations avoid this problem by using the translation "at its rising," leaving "in the east" as an alternative.

The biblical text, as translated traditionally, was so ambiguous in its reference to the east that this brief statement only serves to confuse the issue of the Star and the Magi. However, in the mid-1970s, David Hughes, a scholar and writer with considerable interest in the Star who wrote an authoritative review of the Star mystery for the journal *Nature*, cast considerable doubt on the traditional interpretation of the phrase in Matthew 2:2. He pointed out that the original Greek text says "en té anatolé"[1] (in the east, in the singular). However, the correct grammar in Greek, if you want to say that an object is in the east, is to use the plural "en tai anatolai"[2] (literally, "in the easts"). Hughes points out that the phrase "en té anatolé" has a special meaning in Greek, that is, it denominates the "heliacal rising" of a star or a planet.

To understand the meaning of "heliacal rising," we must remember that all stars, except the ones that are close to one of the poles of the sky, are invisible at certain times of the year. This is because the Sun passes either in front or nearly in front of them (if they are in the ecliptic), blocking out the star completely while it passes behind the Sun. Alternatively—if a star is farther north or south of the ecliptic—at certain times of year it will rise or set at the same time as the Sun. This means the star is only above the horizon during the hours of daylight and cannot be seen at all at night. More technically, from the time the star sets at sunset, to the time, weeks or months later, when it rises at sunrise, it is at *conjunction* with the Sun.

After a certain period of invisibility, the star will become visible again in the morning sky, at dawn. A brilliant star such as Sirius will become visible, low in the bright dawn horizon, as soon as it has separated minimally from the Sun in the sky. This first visibility of a star in the light of dawn is known as the heliacal rising. This phenomenon was important in biblical times because it could be used as an accurate calendar, marking the passing of the seasons. A practical example is the heliacal rising of Sirius, which was used to predict the date when the Nile would flood: when Sirius was first sighted in the dawn sky, it meant that the flood season was imminent. Such observations were valuable to the inhabitants of the Nile Delta, who would then also know when to plant their crops.

According to Hughes, when the Magi stated that they had seen the Star "en té anatolé," they were saying that they had seen the Star in the first light of dawn. This translation has now been adopted to some degree in many recent versions of the Bible, as reflected by the phrase used in the New Revised Standard Version, "at its rising." Thus, as dawn broke in the east, the Magi would in fact have seen a bright star, as most people like to believe. However, this brief comment actually gives us far more information. It tells us that the Star was probably positioned low in the east, at dawn, and this apparently insignificant detail is extremely useful. Why? Because if we can identify the date on which the Star was first seen, we can limit its position in the sky to a narrow band to the west of the Sun, and if we can estimate its position in the sky, we are slightly closer to identifying what the Star could have been. Even the slightly different reading given by the New Revised Standard Version is significant in this sense, as it implies that the Star was quite low in the east or southeast when first seen, although not necessarily in the dawn sky.

Nowhere in the Bible nor in the other texts does it state that the Star went before the Magi in their journey to Jerusalem; this comment is made only with reference to their journey from Jerusalem to Bethlehem. If the Star was seen in the east, at dawn, it is physically impossible for it to have appeared to "go before them" unless they came from a totally unexpected and highly unlikely region. The reason is simple: if the Magi traveled from

Babylon, Persia, or any other state in the east, the Star would have been behind them on their journey. They were traveling toward the west, so the sunset would have been in front of them and the sunrise would have been in the direction from which they came. This, of course, makes sense. The Magi, as we will learn more fully later, had been waiting for a sign, and once they saw it, they knew that they had to travel west to Jerusalem, as that was the capital of the Jewish world.

Suppose we now take the biblical account at face value and accept that the Star did go before them and guided them as they traveled to Bethlehem from Jerusalem. Where must the Star have been in the sky if it were to do this? Bethlehem is just a few miles away from and almost exactly due south of Jerusalem. If the Star really did go before them, it was not in the east, it was in the southern sky at the time. In the time between the first sighting by the Magi and the arrival in Jerusalem, the Star had therefore swung around from the east to the south of the sky. This fact has been used by some experts to suggest that the Star was, in fact, a comet, because only a comet, of all the natural objects, could move this way. Taking into account this important detail of the change in the position of the Star and combining it with the heliacal rising and biblical and other texts, we know a few things about the Star of Bethlehem that will help us in our search for an explanation.

The Star, whatever it was, must have been important, at least to the Magi. Although it is likely that the Protoevangelium of James and the letter of Ignatius may both have exaggerated its brightness to some degree (probably a great deal in the case of the Protoevangelium), the object was in some way unique and drew at least some people's attention. It was sufficiently obvious and spectacular that, as we will see, the Magi knew which of the whole series of astronomical events occurring around the time of the Nativity was the one that referred to the imminent birth of Jesus. Perhaps the Star was not obvious to everybody; King Herod, as the Gospels imply, did not know of it, possibly because it was not deemed important enough to be brought to his attention.

Whatever the details, the Star must have been prominent in some way. This means that it was probably observed by other people, even if they did not realize its significance or, rather, did not interpret it the same way as the Magi. It is probable that the Star of Bethlehem has long been staring us in the face in the form, for example, of ancient Chinese astronomical records, but we have just not been capable of seeing it yet. We will understand why more clearly later in the book.

Another factor to consider is that the Star, whatever it was, must have been visible for a considerable period of time. We know that the journey of the Magi was a long one. If the Magi did indeed come from Babylonia, as is tacitly accepted by many people, then the distance they would have had to travel to get to Jerusalem was around 550 miles, even if they went as the crow flies. One did not make such a long journey in a few hours in those days. The journey would have been made in a camel train or, possibly, on horseback. In either case, to avoid the heat of day, they would probably have traveled largely early in the morning, around sunrise. If the camel train traveled at around two miles per hour and for around eight hours per day, the journey would have taken close to a month and a half. Hence, a star that showed up in the east, at its heliacal rising, could well have moved to the south by the time the Magi entered Bethlehem.

The Magi might possibly have come from much farther away than Babylonia. Persia, an alternative suggested by some experts, would imply a journey nearly twice as long and far more arduous. While one could imagine that a forced march might permit the Magi to arrive at Jerusalem from Babylonia far more rapidly than we have supposed, it is hard to imagine that they could arrive from Persia in anything less than several weeks. To add to the difficulties of the desert, there were several rivers to be crossed (the Jordan, at least, and possibly the Tigris and Euphrates), each of which would have required the location and hiring of a boat large enough to transport them (see fig. 2.1).

It also is unlikely that the Magi would have been prepared for an immediate departure to the Holy Land. They would have seen the Star, pondered—and possibly argued—about its significance, and only then made plans and prepared supplies for the journey. Only a fool would rush into a journey across one of the most inhospitable deserts of the planet without adequate supplies and preparation, and the Magi were certainly no fools. An optimistic estimate is that between planning the journey and carrying it out, at least two months would elapse between the first sighting of the Star and the Magi's arrival in Jerusalem. It is possible that the journey would have taken even longer to complete from the first sighting of the Star. Only if the Magi had been in a desperate hurry—rapidly preparing for departure and driving their mounts hard, could the time between first sighting the Star and arriving in Jerusalem be much less than a month.

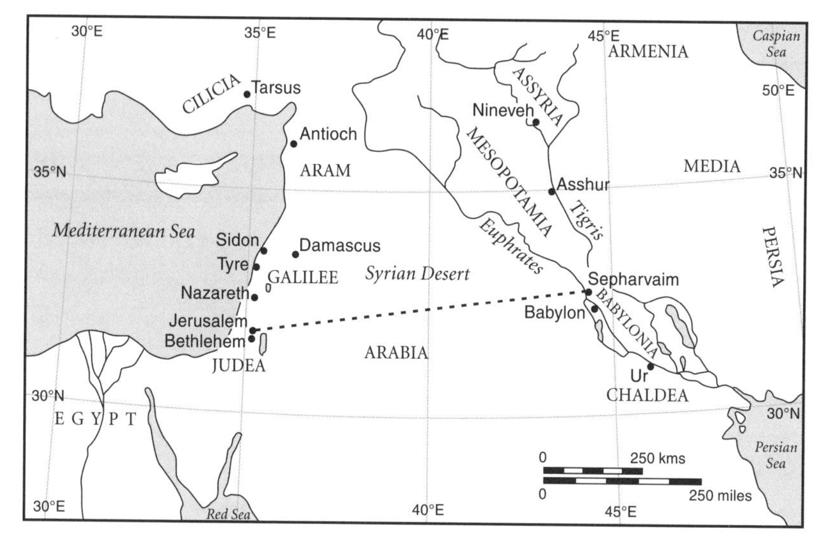

Figure 2.1. Map of region. (Illustration by Chris Brest.)

There is a good reason to suspect that the time spent between leaving on the journey and arriving in Bethlehem was significantly longer than might be imagined: the Magi most probably did not go to Bethlehem directly. We can speculate as to the reasons why. Even if they were royal ambassadors, it is not clear that Herod would have seen them straight away. Perhaps they would also have wanted to rest from their long desert journey first, even before seeking an audience with the king. And then, of course, there would have been the final phase of the journey, a six-mile journey from Jerusalem to the tiny village of Bethlehem.

Matthew 2:3–7 gives us another reason to believe that the Magi spent some days in Jerusalem first. Matthew describes Herod's reaction to the Magi's news in some detail, pointing out that King Herod convened all the scribes and priests of Jerusalem to obtain their opinion and interpretation of events. Only on hearing the results of the deliberations of the conclave did Herod call the Magi to a second audience, at which he requested further information and details about the Star and its appearance. The fact that Herod required such detailed information is clear evidence that he had not personally seen the Star. There are various possible reasons. One is simply that the significance of the Star was obvious only to a scholar or a rabbi, someone steeped in the traditions and beliefs of the Jewish people. Perhaps his courtiers had seen the Star, or knew of it, but did not bother their king with the news because they did not realize its significance. It is possible that the appearance of the Star coincided with bad weather, and nobody in Jerusalem had seen it.

Herod the Great was not a pleasant man. He was effectively a puppet leader who had close ties to the Roman authorities and governed only with their acquiescence and while his actions met their approval. Judea was a protectorate of Rome, which meant it had only as much independence as Rome was willing to give it. A shrewd tyrant, he ruled from 37 to 4 B.C. and even ordered that his wife, Miramme, be murdered to consolidate his position. Perhaps Herod had not heard of the Star because he was being "protected" from news likely to provoke his kingly wrath. It would take a brave person to inform him of the sighting of a new star, one that implied a new king had been born in Jerusalem. From a courtier's point of view, it would have been much safer to let some visiting dignitary break the news. "Killing the messenger" was not merely a literary phrase in those days, but often a fact, particularly in Herod's court.

The implication of the passages in the Bible is therefore clear: the Magi were delayed for some time in Jerusalem while certain protocols, such as audiences with the king, convening of the scribes, and deliberations, took place. All of this would add significantly to their journey time. It is likely that the Star, whatever its origin, was visible for at least two months: Matthew 2:9–10 states quite explicitly that the Star was visible both on leaving Jerusalem and on arriving in Bethlehem, and it had also been visible to them before they started on their journey. As we shall see, this limits probable explanations a great deal, for short-lived phenomena are immediately ruled out.

Some people read Matthew's text as suggesting that the Star appeared, was visible for some time, was lost from view, and finally reappeared again to guide the Magi to Bethlehem. This explanation would tremendously limit the possible explanations, for few astronomical objects appear and disappear in such a way. One can reach such a conclusion only by reading closely between the lines in Matthew. This view subtly suggests that the Magi *had* seen the Star in the east, and then, on leaving Jerusalem, the Star they *had* seen in the east now went before them.

Another way of reading the text is to assume that during the time that had passed between the first sighting and the journey to Bethlehem, the Star had moved from the east to the south, not that it had vanished and later reappeared. This suggestion has often been cited as proof that the Star was either a comet or a planet or planetary conjunction. A planet will, over the course of weeks and months, cross the sky from east to west; a comet may cross the same constellations in just a few nights.

The text of Matthew does give another interesting clue that no one seems to have noticed before. When the Star was first seen, the biblical text suggests, it was low in the dawn sky—in other words, due east, or quite close to due east—but by the time the Magi had arrived in Jerusalem and had their audience with Herod, the Star was in the south. This observation is important because such movement is entirely logical and consistent with thinking of the Star of Bethlehem as a celestial object having a simple astronomical explanation. A star will appear to rise about one hour earlier every two weeks due to the motion of the Earth in its orbit about the Sun. In one month the Earth moves through one-twelfth of its orbit, or an angle of $360/12 = 30$ degrees about the Sun; that is, we see the stars advance, relative to the Sun, by 30 degrees every month. The Earth also rotates on its axis 30 degrees in one-twelfth of a day, or in two hours; that is, it takes two hours of rotation of the Earth to return the stars to the position they held one month earlier (see fig. 2.2).

Suppose then, that the Magi, on leaving Babylonia, saw the Star in the east at dawn. Three months later, at the same time, at dawn, the Star would have moved 3×30, or 90 degrees, in the sky. Instead of being due east at dawn, it would now be due south at the same hour. The account by Matthew would thus be entirely logical if the Magi took around three months to arrive in Bethlehem after seeing the Star for the first time. If the Star had appeared a significant distance south of the celestial equator, for example, it could move from its heliacal rising to being due south at dawn in just two months—the farther south a star is in the sky, the more its heliacal rising is delayed. A star well south of the equator will be first seen

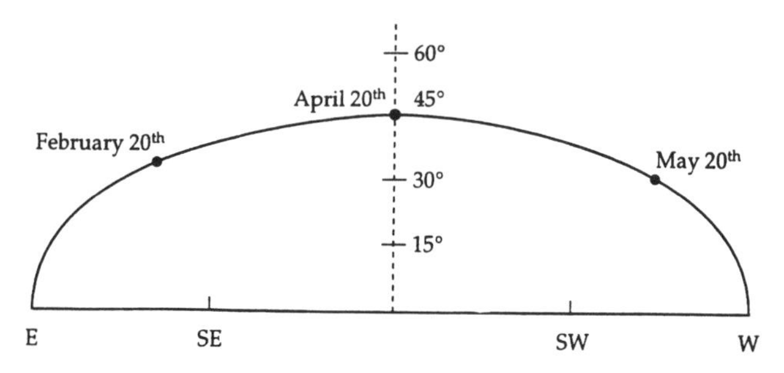

Figure 2.2. Diagram illustrating the change in position of the Star at dawn in three months. (Ramon Castro. Redrawn by Chris Brest.)

toward the southeast. This, as we will see later, has important implications for the explanation of the Star of Bethlehem. The argument does not *prove* that the Star was a simple astronomical event such as a star or nova. What it does demonstrate, satisfactorily, is that the simple details given in Matthew's Gospel do not contradict an astronomical explanation and are, in fact, totally consistent with one.

David Hughes has suggested that the account in Matthew demonstrates that the Star was in the zenith, for only an object in the zenith can appear to stand over a town, or a village, or a building within a town or village. This is an interesting point, but not necessarily conclusive or exclusive. If the Magi approached Bethlehem from the north (as they must have done) and if the Star was in the south (as Matthew's Gospel indicates), there is no reason why Matthew's words that the Star "stood over where the young child was" would indicate that it was in or close to the zenith. It merely indicated that it was quite high in the sky to the south.

Now let us turn to at least one legend of uncertain origin. In the early years of the twentieth century, another astronomer and astrohistorian, E. W. Maunder, wrote what is still a much-referenced book entitled *The Astronomy of the Bible.* Maunder was somewhat polyglot for he had also dedicated himself to the study of the sunspot cycle on the Sun and, by studying three hundred years of detailed observations, had discovered the now famous period of time when sunspots virtually ceased to appear. This lack of major sunspot activity, which occurred from 1645 to 1715, is now rightly named after him—the Maunder Minimum—and represents an important, if little understood fact in solar astronomy.

Maunder's book on the Bible is still regarded as one of the definitive texts about the Star of Bethlehem. In it, he recounts a legend that the Magi, having lost sight of the Star on arriving at Bethlehem, wandered, desolate, around the town. Finally, one of the Magi went to a well to draw water and, on looking down into the well, saw the Star reflected in the water at midday. This legend is both famous and curious and has caused a great deal of debate. Those who believe it use it to demonstrate that the Star was exactly in the zenith and thus "stood over Bethlehem." There has even been a long scientific debate about the practicability of observing stars in daylight from the bottom of wells. Those who favor the legend claim that the darkness caused by the enclosure of the well permits the stars to be seen—one of those curious tales that recur often as so-called common knowledge. Even though the well is dark if sunlight doesn't enter directly, this darkness doesn't affect the brightness of the sky, and you don't have to climb down a well to discover this. Anyone who has gone into a cave in daylight will have experienced the same "well effect": when you look out at the sky from deep inside a cave, the light from the cave entrance is so blinding in contrast to the darkness inside that you can't see anything at all, especially not stars. The same is true for wells.

Some people have even gone to the extreme of testing this theory. One of the best possible places to do so is from the bottom of the Homestake Gold Mine in South Dakota. The mine is 4,850 feet deep. A few years ago, the British astronomer, broadcaster, and popularizer of astronomy, Patrick Moore, tried the experiment of looking directly up the shaft from the bottom of the mine to see if he could actually see stars in daylight. He found, as he fully expected to, that all he could see was a bright blue circle of sky, with no stars at all.

It is true that some celestial objects are visible in the sky in broad daylight, but only extremely brilliant ones. People are often surprised when they see the Moon in the daytime sky, for example. Venus, also, can be seen easily if you know where to look, especially if you stand in the shade to avoid sunlight. Occasionally there are reports that Jupiter has been sighted in daylight, too, in transparent skies, by people with good eyesight. Just occasionally a particularly brilliant comet might be visible in daylight as well. For the Star of Bethlehem to have been seen reflected in the water of a well, though, it would have had to be tremendously bright and clearly visible to the naked eye.

Between the various texts that mention the Star and the myths, legends, and speculations about it, we are left with a series of difficult and, in some cases, intractable questions. If we are to explain the mystery of the Star of Bethlehem, we must accept that some questions are not answerable, and we must concentrate on those that are. The situation is still difficult even if we try to reduce the possible explanations to just a few key questions: When was Jesus born? Where did the Magi come from? Was the Star of Bethlehem seen outside the general area of the Holy Land and Babylonia? For how long was the Star visible?

To solve the mystery, we need answers to all these questions. It has taken us almost two thousand years to get them.

3

The First Christmas

One of the most fundamental questions we must answer is: When exactly was Jesus born? Was he truly born on Christmas Day? If not, why do we celebrate his birth on December 25? Most Christians have wondered about these questions at some time in their lives.

In the Western world, everyone is familiar with the way of reckoning dates as A.D., for "Anno Domini," which means "in the year of our lord" in Latin, and "B.C., which stands for "Before Christ." At first glance, this appears to mean that if the year is A.D. 2000, then Jesus was born 2,000 years ago. Such a simple supposition hides a fascinating story. It turns out that Jesus was not born in the year that our calendar would have us believe.

To find out the date of Jesus' birth—the natural starting point in any search for the Star of Bethlehem—we have to examine the origins of the calendrical system that we use. The accuracy of our calendar is, as we will see, open to considerable doubt.

Let's go back to the days of the Roman Empire. In its heyday, it dominated a large part of the world, although we should not forget the important civilizations in Asia

(China, India, Japan, Korea) that were unknown to the Romans and had little or no contact with the Western world. However, for many people in Europe, North Africa, and western Asia, the Roman Empire and its customs were the ones they knew and adopted—sometimes willingly, sometimes with a little persuasion from Caesar's legions. This meant that the Roman way of reckoning time would automatically be used by millions of Roman citizens and by the inhabitants of the large swath of territory under the direct influence of Rome.

The Romans created a calendar that served to highlight their dominance of the known world, counting the years in terms of Anno Urbis Conditae (A.U.C.), or the number of years that had passed since the founding of Rome. The initial Roman calendar, with its ten-month year and complicated system of leap months, was tremendously clumsy. It contrasted greatly with the Babylonian calendar, still used today by Moslems and Jews, which is based on the lunar year comprised of twelve lunar months (of twenty-nine and a half days each) onto which a leap month is added every third year to take care of the fact that the lunar year of twelve months is eleven days shorter than the solar year. The Roman year started with a month named after Janus, the two-headed god of gates who looked forward and backward at the same time. It continued with the second month for the Roman feast of purification, Februa. The third month was named after the god Mars. April stems from the Latin word *apero*, meaning "the latter" or "second," as it was the second month of the Roman year (March being the first). May came from Maia, the Goddess of increase; June from Juno, the sister (and wife!) of Jupiter; July from Julius;

August from Augustus Caesar. Finally the Roman equivalents were used for *Sept*ember (the seventh month), *Octo*ber (the eighth month), *Nov*ember (the ninth month), and *Dec*ember (the tenth month).

But in 46 B.C., a decree from Julius Caesar changed the whole basis of the Roman calendar, adding two months in the middle of the year. One became known as July (for Julius Caesar), the other, eventually, as August (for Caesar Augustus). These extra two months are why, for example, December, whose name means, as we have seen, "the tenth month," is now the twelfth month of the Western calendar. Caesar's decree gave rise to the modern calendar of twelve months each composed of thirty-something days. He was well advised on these matters by Sosigenes, a Greek astronomer about whom virtually nothing is known apart from his involvement in this reform. In addition to reorganizing the calendar, Sosigenes also introduced the leap year in a stunningly accurate form, which worked successfully for more than sixteen hundred years and—with only very slight modifications—still works today.

This revised calendar, which is the basis for our modern one, became known as the Julian calendar. The year had 365 days, and a leap year every four years added a day. The Roman year began in springtime, with its promise of new warmth, growth, and hope, on the day we would now call March 25. This more accurate Julian calendar proved astonishingly durable, for it remained in place until 1582, when Pope Gregory XIII had a small, cumulative error in the date adjusted. By having a leap year every four years, the Julian calendar had slightly overcorrected the calendar, leaving too many extra days over time. Though the difference is tiny—only eight days per millennium—this small quantity adds up over the centuries and causes the calendar to lose its correspondence with the seasons. By the time of Pope Gregory, the year was off by ten days. To correct this, two things were done: first, the year when the calendar was modified was shortened by ten days and jumped directly from October 4 straight to October 15; second, only the centuries divisible by four hundred are now leap years: 1900 was thus not a leap year, but 2000 will be.

With the Gregorian calendar, the error has been reduced to merely one day every thirty-three hundred years, and even this inaccuracy could be corrected by turning millennium years (e.g., 2000, 3000, 4000) into leap years only if they are divisible by 4000. In other words, for a really precise calendar, the year 2000 should not be a leap year, nor should the year 3000, but 4000 should.

The Gregorian reform of the calendar had some curious social consequences back in the sixteenth century. For example, in some countries people rioted because they thought they had been robbed of more than a week of their lives. The United Kingdom waited almost two hundred years to adopt the Gregorian calendar. It finally accepted the inevitable in 1752 and adjusted the new calendar by eleven days, with September 14 following September 2. The most curious consequence, though, occurs in the British fiscal year: it had ended on March 25, following the Roman calendar, and now it ends on April 5 (March 25 plus eleven lost days), a hybrid date which, no doubt, puzzles many who encounter it.

What are the consequences of such calendar making and recalibrating? When an ancient document says that something happened on, say, December 25, 2 B.C., it is not the same as our current December 25. The tradition is to convert these old dates from their original form (the "Old Style") to the modern, Gregorian date, but this conversion obviously cannot be taken into account in the original documents.

In our modern reckoning, Rome was founded in the year 753 B.C. While the Roman Empire flourished and dominated the Western world, this basis for the calendar was a perfectly acceptable and logical one for the majority of European peoples. Even so, alternative calendars did exist then as now, even within the Western world. In the Jewish calendar, for example, which is thought to have been introduced in the second century B.C. by Yose ben Halafta, the year 3761 corresponds to the year A.D. 1. With the fall of the Roman Empire at the start of the fifth century and when the city was sacked, counting the number of years since the founding of Rome ceased to be the natural way of marking the passage of time for most of the peoples of Europe.

Our modern calendar dates from approximately a century after the fall of the Roman Empire. Both the year and the date of Christmas were fixed by Dionysius Exiguus in A.D. 525. Dionysius was a Scythian monk and church scholar who lived in post-imperial Rome. He chose the name "Exiguus," which means "the little one," as a mark of humility, to distinguish himself from a more famous Dionysius who had lived some two centuries before him and had been an influential writer in the early Christian church.

The calendar, as it stood, was inextricably linked to the pagan practices of the old Roman Empire. There was a need to cleanse it, removing the pagan references and making it compatible with the fast-growing Christian church, which was, even then, beginning to dominate in Europe. Dionysius was called upon to construct a new Easter cycle for the church. Easter is the foremost event in the church year and, according to Thomas Bokenkotter in his *Concise History of the Catholic Church*, it was the first feast to be celebrated churchwide on the same day. For Dionysius to determine the date for Easter was, in effect, to ask him to construct a new calendar around the old one.

To set his calendar apart from its Roman predecessor, Dionysius decided to base it on the date of the birth of Jesus. In so doing, Dionysius ended the Era of Diocletian, as it came to be known, and got rid of the name of this "impious persecutor," as Dionysius described him. This act clearly established the sought-after Christianization. First, though, he had to calculate for himself the date of the Nativity. To do this, he followed precedent and used one of the best and most reliable measurements of time available to him: the reigns of the Roman emperors. By summing up the lengths of their reigns and calculating backwards, he was able to fix the date of the birth of Jesus by what appeared to be a most satisfactory method.

Unfortunately, Dionysius committed two serious errors in his calculation or, at least, two errors we know about. First, he omitted the year zero. Our calendar jumps from 1 B.C. to A.D. 1 without the year 0 in between—a Y0K problem! We thus leap straight from one year before Christ to one year after and lose a year.

The second mistake was more serious. In general, the method of counting backwards the years of reign of the Roman emperors is both excellent and reliable. But Dionysius Exiguus did not go through the process correctly.

What's more, it took more than one thousand years for someone to discover this mistake. In 1605, a Pole named Laurentius Suslyga published a work pointing out that Dionysius's calendar was off by four years. Laurentius had the right idea, but unfortunately he was only four-fifths right: Dionysius was off not by four years, but by a total of five; he either forgot or for some reason excluded a four-year period during which Caesar Augustus reigned under his given name of Octavian, which he used for only a short part of his reign. Augustus was the emperor at the time of the birth of Jesus, so this error is not trivial.[1]

Thus Dionysius Exiguus made at least two mistakes when he calculated Jesus' birthdate, adding up to a five-year error in total. There is no direct evidence that he made any more errors. All that we can say for certain is that Jesus was not born in the (nonexistent) year 0. If we assume that Dionysius made no other mistakes, Jesus would have been born in 5 B.C.—paradoxically, five years "Before Christ"!

Can we say for sure that the date of the Nativity really was 5 B.C.? The answer has to be "no" simply because we do not know if there are further errors in Dionysius's calculations. If there are more, we don't know whether they should be added to or subtracted from our starting point of 5 B.C. We can only say that Jesus' year of birth was 5 B.C. plus or minus a few years. To get a more precise date and to check Dionysius's calculations, we have to look for more clues as to the exact date of the Nativity. If we found that 5 B.C. is the most probable date, then Dionysius Exiguus was correct in his reckoning, apart from the two known mistakes. In fact, some researchers have already concluded that his chronology is completely accurate, once the five-year error is taken into account, and that the Nativity really did happen in 5 B.C. However, other researchers beg to differ.

In their introduction to the New Testament, Norman Perrin and Dennis Duling sum up the problem of Jesus' birth year this way: "Sixth-century A.D. calculations of the birth of Jesus were in error by at least four years. Modern scholars usually date his birth at 6–4 B.C."[2] Such a conclusion can be reached through historical information that gives us a number of excellent clues to when Jesus was born. But, in the end, they merely narrow the range of possible dates, in part because some of the clues contradict one another.

Astronomy gives us one of the key clues in this particular detective story. The Bible tells us that Jesus was born during the reign of King Herod the Great. Contemporary histories report that the death of King Herod occurred shortly after an eclipse of the Moon visible from Jericho, but before the feast of Passover. In other words, he died during a period of a lunar month (twenty-nine days) between the two events. This fortunate coincidence allows us to date Herod's death quite accurately.

Because the feast of Passover is calculated from the Jewish lunar year, which is not the same as our solar year, it can fall during a wide range of dates in March and April (see table 3.1). Passover starts on the fourteenth day of the month of Nisan, which is one day before Full Moon, as there are fifteen days from the start of a lunar month to the date of the month's Full Moon. The Jewish year is calculated from Rosh Hashanah, which usually falls in September, with Nisan being the seventh month.

Passover can thus be easily calculated for past, present,

TABLE 3.1 The Date of Passover between 10 B.C. and A.D. 1

Hebrew Year	*Date of 15 Nisan*	*Gregorian Year*
3750	March 27	11 B.C.
3751	April 16	10 B.C.
3752	April 4	9 B.C.
3753	March 25	8 B.C.
3754	April 12	7 B.C.
3755	April 1	6 B.C.
3756	March 21	5 B.C.
3757	April 10	4 B.C.
3758	March 29	3 B.C.
3759	March 18	2 B.C.
3760	April 6	1 B.C.
3761	March 27	A.D. 1

or future years, and so we know exactly on what date it fell in any year around the Nativity. Susan Stolovy, an astronomer at the Steward Observatory in Tucson, Arizona, calculated the date of Passover for me for a wide range of dates around the Nativity. She writes as follows: "The first day of Passover is always the 15th of Nisan (starting the evening of the 14th of Nisan). These are the Gregorian dates for Nisan 15. Note that the Jewish calendar has a 19-year cycle, in which the years 0, 3, 6, 8, 11, 14, and 17 all have an extra month of Adar (30 days) which occurs in February or March. I happened to notice that the 0th year of one of these cycles is the year 02 (which is 3762), so you can see why [in] some years Passover is considerably later than others." Because we know that the eclipse seen before Herod's death must have occurred less than a lunar month (twenty-nine days) before

Passover (otherwise, a normal, uneclipsed Full Moon would have preceded his death), the next step in the process of establishing our timetable is to identify which eclipse, or eclipses, visible from Jericho occurred less than a lunar month before Passover.

Although lunar eclipses are quite common, they are not *that* common. In one year, at some particular point on Earth, three lunar eclipses may be visible; yet during the same year, at a different place, you might not see a single one. For example, in 1996 two total eclipses of the Moon were seen from the island of Tenerife, where I am writing these words, but the previous total eclipse visible from here happened three years earlier, in 1993. Such spacing shows that it is not difficult to decide which lunar eclipse was the one that preceded the death of King Herod.

Jericho is on the west bank of the Jordan, only 15 miles northeast of Jerusalem, so any eclipse visible from the one city must have been visible (weather permitting) in the other. Computer calculations show that only eight eclipses of the Moon were visible from Jerusalem between 9 and 1 B.C. (see table 3.2). Which of these was Herod's eclipse? Compare the dates of Passover in the table with the dates of lunar eclipses visible from Jerusalem around this time. Only two eclipses in this interval fall around the time of Passover and, of these, the March 23, 5 B.C., eclipse was just after Passover and thus can be discarded. The only eclipse that occurred less than a lunar month before Passover is that of March 13, 4 B.C. No other eclipse occurred within three months of Passover during this period of time, so there is little danger of confusion. Unless there is a gross error somewhere, King Herod must have died in late March or early April of 4 B.C. This

TABLE 3.2 Eclipses of the Moon Visible from Jerusalem between 10 B.C. and 1 B.C.

Date	*Year*	*Type of Eclipse*
June 3	9 B.C.	Total
November 28	9 B.C.	Total
November 18	8 B.C.	Partial, 43%
March 23	5 B.C.	Total
September 15	5 B.C.	Total
March 13	4 B.C.	Partial, 35%
July 17	2 B.C.	Partial, 81%
January 9	1 B.C.	Total

conclusion is regarded by most historians and biblical scholars as an almost certainly correct one. This conclusion is accepted by the New Revised Standard Version (NRSV) of the Bible, which gives in its appendices a chronology of New Testament rulers of Palestine: Roman emperors, Herodian rulers, and procurators of Judea. The NRSV lists the dates of Herod's reign as 37–4 B.C. After his death, Herod's kingdom was divided in three. In Judea, Archelaus, who became ethnarch of Judea, succeeded Herod.

The March 4 B.C. eclipse was not a particularly spectacular one as seen from Jerusalem. Although it began well after midnight, local time, it actually started a little before midnight of March 12 Universal Time (the successor to the old Greenwich Mean Time), hence it is usually referred to as the eclipse of March 12–13, 4 B.C. Most of the city would have been asleep for several hours. Only the guards, watchmen, and shepherds would have noticed how, around twenty minutes before midnight (twenty minutes to two A.M. in Jerusalem), a faint shadow would have started to appear, ghostlike, at the top left-hand side, the northeast, of the Moon's disk; this was Earth's faint penumbral shadow. At 1:45 A.M. local time, the umbra, the dark part of the Earth's shadow, would have touched the Moon and, within a few minutes, the bite out of the Moon's disk would have been obvious. At this time the Moon would have been high in the south-west sky.

On this occasion the center of the Earth's shadow passed somewhat north of the Moon, and the eclipse would never become total from anywhere on the planet. At 2:53 A.M. local time in Jerusalem, the eclipse would have reached its maximum, with 35 percent of the disk of the Moon hidden from the sleeping city. At this point, the top third of the Moon was covered, the shadow covering the Mare Imbrium, Mare Frigoris, and part of the Mare Serenitatis, three of the great lunar "seas." King Herod's men would have seen the eclipsed Moon in the constellation of Virgo, some degrees west of the bright star Spica.

As the Moon started to set, the Earth's shadow would have started to withdraw. By 4 A.M., Jerusalem time, only the faintest sliver of shadow would be left to the north-west of the disk; by 4:01 A.M. only the faint penumbra would still be visible, and even this would remain evident for only a few more minutes. Rarely is the penumbra visible for more than ten minutes before or after an eclipse of the Moon.

Thus, thanks to this eclipse, we know the date of King Herod's death with some precision. If the eclipse happened on March 12–13, 4 B.C., and Passover started on April 11 of that year, the only possible conclusion is the one we have reached: that Herod died in late March or early April 4 of B.C. Given that Matthew describes Herod and his reaction to the news of the birth of Jesus in some detail, we can be certain that the Nativity occurred before his death. To find out how much earlier, we need further clues.

Matthew also supplies us with some more interesting, if somewhat terrifying information. In Matthew 2:16 we read of King Herod's reaction when the Magi had failed to return to Jerusalem and thus did not inform him of the whereabouts of the baby Jesus, the new king of Israel. Shortly after, Herod, in his fury, ordered the killing of all boys aged two years and younger in Bethlehem and its surroundings, in order to make sure that Jesus would be included. In combating a perceived threat to his power, Herod was as thorough as he was cold-blooded, ensuring that there was no chance of escape for his perceived dangerous rival. It is most likely that the order to kill all boys age two and under meant that Jesus was considerably younger than two years at the time of the order. To Herod an extra year or so would ensure a particularly gruesome margin of security. It is also most unlikely that Herod would have thought long about taking this particular step: he would certainly have taken less than a year, and probably just weeks or days, to realize that the Magi would not return. But before his orders could be carried out, Mary, Joseph, and Jesus had fled to Egypt.

Matthew 2:19 then speaks of the death of Herod. The implication is that this followed fairly closely the slaughter of the innocents. If Jesus were one year old at the time of the slaughter—in the Gospels the Greek word used by Herod to describe the child means "toddler," not "baby"—

and Herod died shortly afterwards, this would imply that the Nativity occurred some time early in the year 5 B.C., the same year that we obtain from the corrected calculations by the scholar Dionysius. Even if we assume that Jesus was actually two years old and that the slaughter preceded Herod's death by a full year, stretching the dates to their limits, the Nativity could not have occurred earlier than 7 B.C. My own preference is for the later date. For Jesus to have been born in 7 B.C. would have required King Herod to be particularly slow in his reactions, something which does not tie in with his otherwise ruthless behavior. Even a 6 B.C. Nativity presents certain problems in this sense, but 5 B.C. seems to fit the evidence quite well.

For more evidence to support this date, we can consider the chronology implied by Luke's Gospel. Luke 2:1–7 describes the birth of Jesus as well as other occurrences basic to the traditional Christmas story.

> 1 In those days a decree went out from Emperor Caesar Augustus that all the world should be registered.
>
> 2 This was the first registration and was taken when Quirinius was governor of Syria.
>
> 3 All went to their own towns to be registered.
>
> 4 Joseph also went from the town of Nazareth in Galilee to Judea, to the city of David called Bethlehem, because he was from the house and family of David.
>
> 5 He went to be registered with Mary, to whom he was engaged and who was expecting a child.
>
> 6 While they were there, the time came for her to deliver her child.

> 7 And she gave birth to her firstborn son and wrapped him in bands of cloth, and laid him in a manger because there was no place for them in the inn.

Notice how various details that are usually combined with Matthew's account in Nativity plays are mentioned here and only here: the inn was full and the newborn child was lain in a manger; Mary and Joseph were only in Bethlehem because of Caesar's decree, usually stated to be about taxation, and both had made a long journey with Mary heavily pregnant.

Verses 1 and 2 offer us a series of historical clues—usually called "contextual clues" by historians—that are quite important in determining the date when these events occurred. The verses include much valuable information that has been confirmed by historical documents, although other details are more confusing. Consider the following:

1. Jesus was born when Caesar Augustus was emperor of Rome.
2. A global census was ordered shortly before Jesus' birth.
3. Quirinius was governor of Syria at the time.

Caesar Augustus had a comparatively long reign covering more than four decades around the time of the birth of Jesus, 27 B.C.–A.D. 14. So this does not exactly narrow the birth of Jesus to a specific date. More exacting, though, is the period of Quirinius's office in Syria. Technically, Quirinius was never actually governor before the death of Herod; only in the year A.D. 6 did he finally accede to the governorship, a date that is clearly far too late, based on the aforementioned data, to have relevance in the Nativity. There is a mistake here somewhere. Luke could well have made an error by writing "governor" instead of "legate."

The British astronomer and astrohistorian David Hughes comments that a possible solution to this problem is the fact that historical documents tell us that Quirinius was an emperor's legate in Syria under the governership of Saturnius between 6 and 5 B.C. This is a period that overlaps the most likely date of the birth of Jesus based on our timing of Herod's death. Provided that we have interpreted Luke correctly, this is a useful confirmation of the suggested chronology. Alternatively, it has been suggested that Luke's passage was incorrectly translated. An alternative translation of Luke's verse 2 is this: "A census took place before the one when Quirinius was governor of Syria." That is, there was a pre-Nativity census, and this census was *like* the one that took place during Quirinius's governorship.

Some authors have suggested, due to this contradiction over Quirinius's governorship, that the birth of Jesus might have occurred much later than is supposed here, perhaps as late as A.D. 2. The fact that Quirinius's first stay in Syria agrees well with the earlier date of 7–5 B.C. seems to rule out a gross error in timing, provided that either Luke committed an error in his text with respect to Quirinius's official title, or that scholars have simply interpreted Luke's text incorrectly, either of which is perfectly plausible.

A most problematic issue remains, however: the census took place before the Nativity, but how long before? Caesar Augustus decreed the taking of three censuses for Roman citizens. He called for one in 28 B.C. another in 8 B.C.

and yet another in A.D. 14. The date of the 8 B.C. census is known because the actual document ordering it was found some years ago in Ankara, Turkey. What can we make of this substantial gap between the date of Caesar Augustus's order in 8 B.C. and the date of Quirinius's legateship in 6–5 B.C.?

The discrepancy can be explained in various possible ways. The first is to suggest that slow communications within the Roman Empire would have made it impossible for the census to be carried out before 5 or 6 B.C., even though the order had been given in 8 B.C. Although some outlying parts of the Roman Empire might have suffered from poor communications, this is not generally in accordance with what we know about Roman infrastructure and efficiency. Successive Roman emperors knew that the empire would not be governable if communications were not fast and effective; therefore enormous efforts were put into constructing the finest roads to expedite good communications.

Syria, though, is on the shores of the Mediterranean, and its principal cities and towns are close to the coast and a few weeks' journey at worst from Rome by sea and road. Even if we assume, as would most likely have been the case, that individual citizens would have traveled very slowly to carry out an unpopular action such as a call to census so that taxes could be collected from them, it is hard to believe that as much as three years, 8–5 B.C., might have been necessary between the order and its fulfillment.

An alternative suggestion is that Augustus made the order as suggested, but allowed generous time for its dissemination and execution. There would have been no problem with making the order in 8 B.C. but stating that it should be executed in, for example, 6 B.C. In this way he would have ensured that there would be no excuses for delays or disorganization in any province, however far from Rome. All citizens would have ample time to learn of the decree and make the necessary arrangements for their journey and for a substantial absence from home.

A third suggestion is that the 8 B.C. census is not the one referred to by Luke at all. This census was for Roman citizens and, as Joseph was not a Roman citizen but rather a citizen of a conquered nation, he would not have been required to register. A further possibility is that the census to which Luke refers was not for tax-gathering purposes but rather to register an oath of loyalty to Caesar Augustus and to obtain a list of all persons suitable for military service. Even as a non-Roman citizen, Joseph would not have been exempt from this duty. If so, this registration would have happened about one to two years before the death of King Herod, in 5 or 6 B.C. It would have served the useful purpose, from a Roman perspective, of making it clear to the Jews that they were subjugated, whatever local autonomy they might enjoy. There is ample precedent for such registrations in Roman history, and Colin Humphreys, a British researcher into the Star of Bethlehem, has suggested that contextual clues support a date for it approximately one year before Herod's death, in other words, in 5 B.C.

Thus the evidence relating to Luke's mention of the census is rather inconclusive. It might have been the 8 B.C. census, but it is just as possible that it was not. Certainly Luke's mention that "the whole world" was called to be registered does not tie in with a Romans-only decree. It has, however, been suggested that the phrase "the whole world" or its alternative translation, "the inhabited earth," was used with hyperbole and was no more than a part of official rhetoric, with no intent that it be taken literally.

We cannot, thus, precisely date the birth of Jesus from the evidence, but we can confine it to a small range of years. It is unlikely to have been later than 5 B.C. because we can date the death of King Herod quite exactly. The majority of the evidence suggests that 5 B.C. was the correct year, although 6 B.C. is possible. An earlier date such as 7 B.C. or a later one such as 4 B.C. looks rather unlikely and is inconsistent with other strong pieces of evidence. Although a date of 7 B.C. ties in better with the known census ordered by Augustus, it creates more problems than it solves. A 5 B.C. Nativity also leads to the rather satisfying conclusion that history vindicates Dionysius as being largely correct, apart from the two known, silly slips which were, to a certain extent, understandable.

If we stick to the principle that the best theory and the one most likely to be correct is the simplest one, we can say that it is quite probable that the birth of Jesus occurred in 5 B.C. The Spanish theologian, Andrés Brito, my good friend and former colleague, who has lectured on a variety of biblical topics including the Star of Bethlehem, is just one expert who agrees that this seems to be the best compromise date between the different ones suggested.

Now that we have reached this conclusion, we can ask the question: "Was Jesus really born on December 25, 5 B.C.?" This traditional date of Christmas has been unchanged for many centuries. In fact, the date appears to have been set almost a century before the fall of Rome. Jack Finegan, an American scholar of New Testament history, in his definitive work *Handbook of Bible Chronology*, states that December 25 has been used since around A.D. 336. Of course, this is only the date of its first mention. The reason for this choice of date is simple: it coincided with the pagan midwinter festival of "Sol Invictus" (the undefeated Sun). This festival was an old one even in Roman times, already being celebrated by the Celts. It marked the winter solstice, the shortest day and longest night of the year. As in modern times, the feeling that the shortest day had occurred and is now past gave rise to a feeling of relief that the worst of winter was finally over, although, astronomically, midwinter actually occurs some six weeks later.

One of the traditional names for Christmas is "Yuletide," from the word yule or laurel, and the name is evocative of the celebrations that were held during this pagan holiday. The details are remarkably similar to those of modern-day Christmas, although different branches of the Christian church highlight different aspects of the old festivities.

The festival was a public holiday in which all nonessential work ceased. Efforts were dedicated more to the decoration of houses than to the labors of the workplace; even here the modern celebration is similar to the old. Today, in accordance with the German tradition exported around the world since the mid-nineteenth century, families decorate their houses with evergreen pine trees (or a reasonable simulation of one) and, in many places, hang sprigs of holly. In Roman times, houses were also decorated with evergreens, although they were more likely to be laurel, or yule. Then, as now, this was a highly symbolic gesture: the bright green branches adorning one's house show that winter cannot destroy *all* things green, however many other trees were reduced to stripped boughs, naked of leaves.

Most surprising to many is the fact that the festival has always been a time for giving and receiving gifts. It is often stated incorrectly that the Christmas tradition of giving and receiving gifts springs from the offerings of the Magi. According to Matthew 2:11,

> 11 On entering the house, they saw the child with Mary his mother, and knelt down and paid him homage. Then, opening their treasure-chests, they offered him gifts of gold, frankincense and myrrh.

In fact, around the time during which Jesus was born, it had already been a long-standing tradition to give gifts at the time of the great midwinter festival. The action of the Magi, described by Matthew, changed it from a pagan ritual to a Christian practice and tradition. By now this tradition has been somewhat transformed, with a variety of personalities bringing presents on different days in the month of the midwinter festival.

Yet another tradition which most people think of as modern was already a traditional part of the old midwinter festival. No modern Christmas is complete without its ration of parties and special meals, and the midwinter festival, too, featured similar events. The consumption of alcohol assumed perhaps a disproportionately important role in the celebrations.

The final part of the traditional midwinter festival may not be familiar to those living in mainly Protestant countries such as the United Kingdom and United States. But it is familiar to those in devout Catholic countries, particularly in Spain and Italy, and in Central and South America: the time is an occasion for parades and processions of all kinds, for celebrations in the streets, for all to watch and to join if they desire.

It is therefore not surprising that, as Christianity extended its hold on the late Roman Empire, the masses had no desire to lose their traditional holiday and festival. Christianity, in its great tradition of compromise and adaptation, chose not to embark on the hopeless task of suppressing such a popular celebration, but instead decided to integrate it into the growing Christian tradition and make it the date of one of the most important events of the Christian calendar. Thus, the adoption of December 25 as the date of the birth of Jesus, in a manner of speaking, just "happened." It has no real basis in history as a suggested date for the Nativity.

Indeed, historical records suggest all manner of different dates as the actual date of the birth of Jesus. The Armenian Apostolic Church celebrates the birth of Jesus on January 6, for example, as does the Greek orthodox. Catholicism also marks this date as the Feast of the Epiphany, which commemorates the visit of the Magi to the infant Jesus, and is the Catholics' traditional date for giving Christmas gifts. Modern commercialism is, however, changing this practice in parts of the world; in Spain, for example, many families give gifts both on December 25 and January 6; in Britain, the royal family follows the German tradition of opening its gifts on December 24; in still other countries, gifts are given on the Feast of Saint Nicholas—the original Santa Claus—on December 6.

There have been many attempts to fix the date of the birth of Jesus, and Jack Finegan notes several fascinating early ones. One of the earliest was made by Hippolytus (ca. A.D. 165–235) first of the so-called anti-Popes, whose statue inscribed with his Easter tables and list of writings, was unearthed in Rome in 1551. According to him, Jesus was born 5502 years after Adam, giving a date of around 2 or 3 B.C. He refined this further by saying that Jesus was born "on the fourth day before the Nones of April." The *nones* are the ninth day before the *ides* (the 15th or 16th day) of any particular month. In April the nones fall on the fifth day, so the exact day of birth would have been April 2.[3] An alternative and quite astonishing calculation, detailed by Finegan, comes from an anonymous North African writer in the work, *De Pascha Computus*. This author says that creation began on a Sunday, in which light and darkness were separated. As the two were equal, the date must have been an equinox, and thus, in the Roman calendar, the date of creation could only have been March 25. The Sun was created on Wednesday and, as Christ is the "sun of righteousness," Jesus must have been born on Wednesday, March 28.

In 194 A.D., Clement of Alexandria fixed the date of the birth of Jesus as November 18, 3 B.C. Even Clement exercised caution and offered two alternative dates: April 19/20 and May 20. Epiphanius (ca. 315–403), bishop of Salamis on the island of Cyprus, writing approximately a century and a half after Clement, fixed the date of birth as January 6 but also gave the date of May 20, stating that this was the date of the Conception. Epiphanius also gave May 21 and June 20 as alternatives. An early January birth is possible if the conception was in late May and the pregnancy did not reach its term (it would have been only seven and a half months for a May 20 conception). This arrangement of dates would have made the child seriously premature and the birth highly problematic. A late June conception and early January birth would have made the newborn baby's survival highly unlikely, particularly under the conditions that Luke's Gospel describes.

Possibly the May date is a simple confusion and refers not to the conception but to the birth of Jesus. During the 1950s the German journalist and author Werner Keller, whose book *The Bible as History* defined the field, pointed out that the description of the Nativity in Luke is not consistent with a winter birth. Luke 2:8–19 speaks of the announcement of the birth to the shepherds and their visit to the child. The key verse is the following:

> 8 In that region there were shepherds living in the fields, keeping watch over their flock by night.

Keller points out that the winter weather in Palestine, though not as severe as at more northerly latitudes, is still unpleasant. During December there are generally heavy rainfalls, and through the entire winter period there are, by turns, rain, frost, and, occasionally, even heavy snowfalls. Note that Bethlehem, in particular, would have a cooler climate than more coastal towns in the region, as it is at the respectable altitude of 2,550 feet above sea level. This would chill its climate quite significantly, particularly without the proximity of the warm sea to counteract the effects of altitude. The snow line would often fall low enough for snow showers to sprinkle the hills, burying the forage. Therefore, at least in modern times, the shepherds of the region, although they are as hardy as ever, would not think of staying out on the hillsides with their flocks through the winter months. At this time of year the sheep are kept under shelter until lambing time approaches, when they are put out to pasture again. The time of year when the attention of the shepherds on a twenty-four-hour basis would be most needed would be March and April. In these months, as lambing occurred, the shepherds would need to be vigilant, protecting the flock from danger and helping ewes in distress. Luke's passage suggests that the birth of Jesus was most likely to have occurred in early spring, around March, April, or perhaps May, at lambing time, or when the newborn lambs would still have required extra attention. Intriguingly, this time might also serve as the origin of the expression "the holy lamb of God," which is used so often to describe Jesus.

If the birth were to have occurred in March or April of the year 5 B.C., then the order of Herod to slaughter all infant boys aged two years or younger would have been a logical one by his standards, if still an overreaction, to the Nativity. These months agree with the suggestion by David Hughes that the massacre was ordered in 5 or 6 B.C.

It has also been suggested that the slaughter was on a much smaller scale than is generally believed because Bethlehem is, and was, a small town. Even when all the surrounding villages are included, the number of babies less than two years old who would have been included in the massacre could not have been large. That said, no matter whether the number of babies murdered out of revenge or self-preservation was a few hundred or "only" a few tens, it marks the type of person that Herod was. History has probably been kinder to him than he deserved.

Some researchers have tried to interpret much farther

the events described in Matthew, chapter 2, and Luke, chapter 2, giving a complete chronology of the events described in these two Gospels. This is somewhat complicated by the fact that the only event in common between the two Gospels is the actual birth of Jesus. Luke describes the decree, the journey of Mary and Joseph, the fact that the inn was full and that Mary had to use a manger as a crib, and the visit of the shepherds. We see the historically minded Luke giving a detailed chronological account of the events that took place, as well as the more prosaic version of the shepherds. Matthew, in contrast, describes the apparition of the Star, the visit of the Magi, their audience with Herod, and their visit to Bethlehem to pay homage. His account has many more "poetic" details, is more mystical and is less of a strict historical version.

The events described in the typical Nativity play are a mixture of scenes from the two Gospels. With few clues as to how to order them into a precise sequence, it is difficult to say, for example, whether the shepherds or the Magi visited first, and if the child was newborn or several days or weeks old at these times. Luke tells us that the shepherds' visit occurred before Jesus' circumcision on the eighth day, but little more.

An attempt to produce a complete chronology of the events in Matthew and Luke risks overinterpreting the little information that is available. Such accusations are, however, inevitable, given that so many millions of words have been written about the few short verses in Matthew, which mention the word "star" exactly four times.

In 1991, British researcher Colin Humphreys suggested the chronology in table 3.3, which should be read in the

TABLE 3.3 Colin Humphreys' Chronology of the Nativity.

Date	*Event*
5 B.C.	
March 9–May 4	Birth of Jesus in Bethlehem
March 9–May 4	Visit of the shepherds
March 16–May 11	Circumcision
April 18–June 13	Presentation at the Temple in Jerusalem and return to Bethlehem
April 20–June 15	Visit of the Magi
Late April–mid-June	Flight to Egypt
4 B.C.	
Late March	Death of Herod

light of the comments made above. He suggests two alternative and more precise chronologies, one of them based on the assumption that the birth occurred during Passover between April 13 and 27, 5 B.C. This would explain Luke's account that the inn was full, since many visitors would have been there for the Passover celebration. Hence, Mary and Joseph would have had no alternative but to use the stable and put the baby Jesus in the manger. (Incidentally, the stable is never mentioned in the Gospels but it has become a part of the Christmas tradition.) With this chronology, the birth of Jesus is fixed in a rather narrow two-week interval of time but is in broad agreement with what we know of, or can guess from, other sources.

Humphreys' second alternative chronology is based on the suggestion that the phrase "the holy lamb of God," used to describe Jesus, may have meant that he was born on the day when the Passover lambs were chosen. If 5 B.C. was the correct year, this ceremony would have been carried out from sunset on April 14 to sunset on April 15, at the start of the Passover festivities. Once again, we are in danger of placing too much emphasis on what are undeniably interesting clues, but for which there is little evidence. A counterargument would also be to state that, had Jesus been born on such a significant date, some record of this event would have occurred elsewhere above and beyond such an ambiguous comment in the Bible.

David Hughes offers a less precise, more conservative chronology in his seminal 1976 review in the scientific journal *Nature,* shown in table 3.4. Note that Humphreys and Hughes disagree as to the exact interpretation of Dionysius's calendar, with Humphreys taking the view that

TABLE 3.4 David Hughes's Published Chronology of the Nativity.

9–6 B.C.	Saturnius, governor of Syria Quirinius, emperor's legate
8 B.C.	Augustus Caesar decrees that all should be taxed
7–5 B.C.	Biblical date of the birth of Jesus
6–5 B.C.	Slaughter of the innocents
4 B.C.	
March 13	Eclipse of the Moon
March 13–April 11	Death of Herod
April 11	Start of Passover
3 B.C.	
November 18	Nativity according to Clement
2 B.C.	
January 6	Nativity according to Epiphanius
A.D. 1	
December 25	Nativity according to Dionysius Exiguus

the Nativity would have occurred on December 25 of 1 B.C. Though many of the elements of the two basic chronologies are similar, there are some important differences of emphasis, leaving us in doubt which of the two years offered is the correct one. If we are unsure of the exact year of the Nativity, it is even more risky to speak of it as occurring on a specific date in a particular year. We may never know the exact answer to this question, although the evidence that the birth took place in March or April of 5 B.C. is really rather strong. The same time of year in 6 B.C. is also possible, but considerably less likely, while a 7 B.C. or 4 B.C. date seems rather improbable in the light of other evidence. It is also interesting to note, however, that, when speaking on British television over the Christmas holidays in 1998, David Hughes expanded significantly on the chronology he published in 1976 and developed in his 1979 book on the Star of Bethlehem: he now believes that Jesus was born on September 15, 7 B.C. We will discover the reasons for his choice of this date in a later chapter.

Not everyone will agree with the March–April 5 B.C. date, although there seems to be a consensus among most experts that it is about right. Some researchers, however, disagree. In 1996, Ernest Martin, a historian and theologian, challenged the standard chronology and suggested that the Nativity occurred in either 3 or 2 B.C., finally settling on the former. A factor that weighs very heavily in his dating is his identification of "Herod's eclipse" as the January 10, 1 B.C., eclipse and not the one observed in 4 B.C. This identification considers the datum that the eclipse took place less than a month before Passover to be incorrect. Certainly, the eclipse referred to by Martin was much more spectacular than the 4 B.C. eclipse. It was total, with totality lasting slightly over an hour and a half, from 1:35 A.M. to 3:06 A.M. local Jerusalem time. The whole eclipse, with the Moon in the constellation of Cancer, began at thirty-seven minutes past midnight and did not end until 4:04 A.M.

It is unlikely that we ever will or can determine the date of Nativity precisely. My strong preference is for 5 B.C. over the others. As we will see in a later chapter, there is one further factor that tilts the balance in favor of the 5 B.C. date. When we weigh this extra factor in, the balance of probability, even though this is still not total proof, tilts the likelihood heavily in favor of March of 5 B.C. If we do accept 5 B.C. as the date of Jesus' birth, the millennium, which so many plan to celebrate (incorrectly) on December 31, 1999, had already occurred in 1995—and no one noticed.[4]

Whatever errors there may be in the calendar, there seems little point in changing it now. Any change would be senseless, as we are as likely to get it wrong again as to make things better. Even if we were able to put an exact day on the Nativity, December 25 has served us well as a date for Christmas for nearly seventeen centuries, so why change it now? The same comment goes for "correcting" Dionysius's five-year error. It would be change simply for the sake of change.

BEFORE we leave the subject of the chronology of the life of Jesus, it is also of interest to speak briefly of his death, for here we can get some additional indirect evidence of his birth.

As we have seen, eclipses, both of the Sun and of the Moon, have become valuable tools for dating historical events largely because they are often recorded along with important human events. It is even possible that an eclipse played a further important role in the story of Jesus, allowing us to fix the date of the crucifixion. Various authors have suggested that the crucifixion occurred on April 3, A.D. 33. Luke states that Jesus started his ministry "when about 30 years old" (Luke 3:23). Assuming Jesus was born in 5 B.C., he would have been thirty around the year 26 A.D. Later in Luke, we find an interesting reference within the story of the crucifixion that may help to fix when it occurred. Shortly before the death of Jesus on the cross, Luke 23:44–45 states that:

> 44 It was now about noon, and darkness came over the whole land until three in the afternoon,
> 45 while the sun's light failed.

The New Revised Standard Version of the Bible offers the alternative "while the sun was eclipsed." Note that a recent Spanish translation of the Bible, which I have used as a source, uses the word "eclipsed" quite specifically, although most English translations are less clear and speak simply of the sky being darkened.

The Apocryphal Gospel of Peter includes the following fascinating comment:

> And many went about with lamps, supposing it was night, and fell down.

But what if this, like the Star, is another example of an astronomical event being incorporated into the Gospels? In that case, might we think of the two verses in Luke as a description of a solar eclipse? At the suggestion of British physicist Trevor Lipscombe, I checked historical eclipses to see if any were visible from Jerusalem around A.D. 30. Total eclipses of the Sun are rare, at least from any one point of the Earth. Witness just how few total eclipses of the Sun in the last two or three decades have been visible from North America, or from Europe, despite the large surface areas of these two regions.

On November 24, A.D. 29, an eclipse was seen which started in the North Sea, just to the west of Denmark, close to the East Frisian Islands. The Moon's shadow then crossed Europe from Germany through to Bulgaria and raced across Turkey. The total eclipse then crossed the Mediterranean, passing over Larnaca in Cyprus, then Lebanon and Syria. Jerusalem was just slightly south of the band where the eclipse would have been seen to be total: even so, the Sun was 95 percent eclipsed in Jerusalem at 9:05 A.M. Universal Time (11:05 A.M. local time, intriguingly close to the noon hour mentioned by Luke), and the eclipse would have been total in Nazareth and Galilee.

If Luke is really describing an eclipse of the Sun, there is almost no alternative date. From A.D. 25 to 35, just five eclipses of the Sun were visible from Jerusalem. None was anywhere near as important as the A.D. 29 eclipse, and only this one would have been visible as a total eclipse from anywhere near Jerusalem.

The eclipse of the Sun would have lasted for a maximum of two minutes and two seconds on its center line, although, around the Sea of Galilee, the total eclipse would have lasted just one minute and forty-nine seconds. Curiously, in Jerusalem, although 95 percent of the Sun was eclipsed, the sky would not have darkened visibly. Due to the great range of intensity that the human eye can adapt to, the sky does not usually appear to darken during a partial eclipse if one is outdoors.

If, however, a person were indoors and thus with a much more limited amount of light, he or she most definitely would have noticed when the Sun was reduced to just one-twentieth of its normal intensity: the interiors of houses would have become very dark indeed. Similarly, if the sky were cloudy—not with light clouds but with heavy, threatening ones, as is quite possible in the month of November—the effect of darkness could have become quite frightening, especially because the sun would not have been visible and people could not have known that the reason for the lack of light was simply an eclipse.

I have three times experienced this curious effect of the sun being hidden without a noticeable drop in light intensity. The first time was in Finland, where I witnessed a total eclipse while perched claustrophobically with some forty other people on the top of a fire watchtower on a low hill in the middle of the beautiful and endless Finnish forests. But there was no noticeable drop in light until the Moon's shadow finally arrived. The second time was in Morocco, where I watched an annular eclipse in which the Moon covers all but a bright ring around the circumference of the Sun. Observing from a beach just outside Casablanca, I did not notice any sensation of darkness, even at maximum eclipse (96 percent), just before sunset. The third time was in February 1998, when a total eclipse passed just north of the Canary Islands, shortly before sunset, and there was no noticeable darkening then, either. In all three cases, though, most of the sky was clear. But if there had been a lot of heavy cloud cover *and* almost total eclipse, I have no doubt that there would have been appreciable darkening, even in the open air.

Unless Calvary was within the narrow strip of land where the total eclipse was visible, there would have been no darkness. Even if it was there, the darkness would have lasted less than two minutes, rather than the several hours Luke's Gospel appears to suggest. Even for those in their homes, the effect of darkening would have lasted not much more than a quarter of an hour, barely enough time for the people to find and light their lamps. The full eclipse, from its start, as the Moon first entered the Sun's disk, through to its end, would have lasted about an hour and a half in total and certainly not the three hours that Luke recorded.

This whole argument might be based on a misunderstanding. Other writers have suggested that the translation "eclipsed" is impossible because it would have implied an eclipse at Passover (Full Moon). But eclipses of the Sun can occur only at New Moon. But if the eclipse of November 24, A.D. 29, did occur around the time that Jesus was crucified, then we have some degree of confirmation: from Luke we know that Jesus was about thirty when his public ministry, which lasted about three years, began. To have died in A.D. 29, at age thirty-three, is to have been born in about 4 or 5 B.C.

This chapter on the first Christmas need not end on so gloomy a note as the crucifixion of Jesus. We are on the search for a Christmas Star, and we now know in what year we have to look for it—if, of course, it ever existed. But our search is only just beginning as there are a multitude of astronomical occurrences to consider, any one of which might have been linked to the Star. These need to be examined and, like farmers, we must weed out our field before we can harvest our crop.

4

Halley's Comet and Other Red Herrings

I WAS BITTEN by the astronomy bug while still a child. Often I would sneak out of bed to look at the sky through the window when my parents thought I was snugly tucked in. Of course, this was a lot easier to do on the cold nights of winter when it gets dark by about 5 P.M. in southern England, where I lived, and when it is still pretty dark at 7 the next morning. In summer, when nightfall came after 10 P.M. and daybreak at 4 A.M., seeing the night sky was a rarity.

The Christmas sky was a particular favorite of mine, especially the dawn sky. On occasions, it was bejeweled with a crescent Moon and the brilliant "star" which is the planet Venus. I can still recall looking at them together in the sky one Christmas morning and marveling at their beauty; looking back with the aid of my computer, this must have been Christmas Day 1967, when I was only seven years old. If my thoughts had turned toward the Star of Bethlehem back then, I, like many others before and after me, would have been convinced by the glorious sight of the Moon together with Venus that my search for *the*

Star was over. Curiously, Venus and the crescent Moon won't be visible again together in the sky on Christmas Eve or on Christmas morning until the year 2092. Three years later, I would have been even more certain. On Christmas Day 1970 the Moon, Venus, Jupiter, and Mars all formed a tight grouping in the dawn sky. But you can't have everything: that particular Christmas morning the sky must have been dull and cloudy from my home, for otherwise I would surely have remembered such a spectacular sight.

Every year, around Christmastime, when the Star of Bethlehem is discussed, certain ideas about its provenance recur again and again. One of the most frequently suggested explanations is that the Star of Bethlehem might be the planet Venus and, having seen it in the sky on that long ago Christmas morning, fairly close to the Moon, I can understand why. In fact, this same suggestion has just been made again in an article in a well-known professional astronomy journal. Another favorite theory points to Halley's Comet. This is by no means a new theory, as we will see later in this chapter when we visit a chapel in Italy built nearly seven hundred years ago.

Astronomers and amateur observers of the Christmas sky have offered other suggestions as well. Some seem far more plausible than others. We shall look at some of them here, but with no pretense that every single theory ever proposed is included. We shall also examine another category of theories. All are serious attempts to explain the Star of Bethlehem, and while each seems to have some flaw when it is examined critically, none of them is in any way absurd. Even if we end up rejecting these theories, they may still hold some grain of truth in them.

Was It Venus?

However special Venus might look in the Christmas sky when you think it might resemble the Star of Bethlehem, it is still rather commonplace. The planet Venus can be seen in the sky around Christmastime in most years—that's why it is so often suggested to be the Christmas Star. Some years Venus is visible in the morning sky, and other years it is seen after sunset. In 1994, for example, Venus was almost at its maximum brightness and maximum distance from the sun in the morning sky. In 1995 it was low in the west after sunset. In 1996, it was again a morning star, hanging low in the east at sunrise on Christmas Day. In 1997 and in 1998, as in 1995, Venus was low in the west after sunset on Christmas Eve.

At the end of the millennium, at Christmas 2000, in contrast to 1997 and 1998, Venus will be high in the sky and visible for some time after sunset. Even more spectacular is the view awaiting those who rise before dawn on Christmas Day in 1999. Neither year, however, will quite match Christmas Eve of 2022, when Venus, Mercury, and a very thin crescent Moon will all be very close together, low in the twilight sky. Many people will still be stunned by the beauty of the planet in these years, quite high in the East. Many will imagine that this is how the Star of Bethlehem must have looked, shining brilliantly for the Magi and, in 1999, will ask themselves whether the Star has appeared once again in the dawn sky. Others will realize what this striking object really is, and they will wonder if the Magi might have reacted the same way.

Venus has been known for thousands of years. The so-

Figure 4.1. The Venus Tablet. (The British Museum.)

called Venus Tablet (see fig. 4.1), which records the first observation of Venus and was made by an unknown Assyrian astronomer, probably dates back to around 1700 B.C. It is on display in the British Museum in London. When we look at it, we are looking back almost to the dawn of astronomy. Venus was known to the Babylonians as Istar (sometimes spelled Ishtar), the mother of the gods and the personification of woman. In the space age, this ancient name has been celebrated by naming the largest plateau on the surface of Venus, discovered and mapped by radar by NASA's orbiting probes, Ishtar Terra. Other ancient observations of Venus are found on the Ninevah tablets of 684 B.C., which are discussed in chapter 7.

Although surprisingly few observations of Venus have survived the centuries and millennia, the planet was also well known to other great civilizations and was undoubtedly discovered in prehistory by the ancestors of modern *Homo sapiens*. For the Chinese, Venus was known as Tai-pe, and the Egyptians called the planet either Ouâiti or Tioumoutiri, depending on whether it was an evening or a morning star. It is probable that many such civilizations had two names for the planet because they genuinely believed they were seeing two different objects.

The modern name "Venus" comes from the Roman goddess of love. In the pretelescopic age, Venus would have appeared to be the most lovely of stars in the sky and would inevitably have been associated with the goddess. The romantic vision of Venus lasted well into modern times. Even in the early 1960s, a significant group of experts still thought that Venus might have surface oceans and swamps. They thought it might resemble Earth as it was hundreds of millions of years ago, during the Carboniferous Age, when the coal seams exploited around the world were laid down. Others offered an alternative view: they saw the planet as a giant dust bowl, dry and arid. This alternative picture was less attractive and, probably because of that, struck less of a chord. Given a choice between picturing Venus as a world of amphibians, dinosaurs, and exotic vegetation, all living in a vast warm swamp, and seeing it as a lifeless version of Tatooine in *Star Wars,* it is no wonder that people preferred the former.

It is appropriate that such a mysterious planet, thickly overlaid with dense clouds, always hiding its surface from our view, should give rise to such romantic ideas. Sadly, the truth is different. Radio telescope observations from Earth, beginning in 1956, gave the first indications that the surface of the planet was a lot hotter than it should be for its distance from the Sun. Nevertheless, the romantics still held out, explaining the result by suggesting that the excess radiation detected from the planet could come from an ionosphere in its upper atmosphere and not necessarily indicate a baked surface. But soon the romantics made a quick retreat as reality caught up with them. In only six years of measurements, Venus was transformed from paradise into an arid, hellish, fiery inferno.

The result that confirmed the arid nature of the surface of Venus was sent back by the first successful interplanetary probe, Mariner II. This probe, which passed by the planet in 1962, demonstrated that it was the *surface* of the planet and not its upper atmosphere that was exceptionally hot. Any lingering hopes that there might be some terrible error were removed by a succession of later space probes, both American and Soviet. After the initial fly-bys, one probe after another was sent into the atmosphere of Venus and, finally, they were able to land on the surface. But the first space probes that "landed" on the planet did not reveal the full, terrible truth: as we now know, they were in fact destroyed high up in the atmosphere and never even came close to the surface.

The final, unappealing blow to the romantics came

when the Soviets landed the probes Venera 9 and Venera 10 on the Venusian surface. Those probes, in one of the most unexpected coups of the space age, successfully sent back photographs of their surroundings and made measurements of the temperature and conditions into which they had settled. The surface of the planet had a temperature of around 450 °C (842 °F), a temperature that would rapidly burn a Thanksgiving or Christmas turkey into a cinder. Later, the Venera 14 probe, which landed in 1982, measured the even higher temperature of 465 °C (870 °F) on the surface of Venus. And as if that were not enough, the atmospheric pressure on the surface of Venus was shown to be so great that the slightest breeze would lift any astronaut into the air and crash him or her against the rocks. With a surface pressure as high as ninety-four times greater than that on the Earth's surface, a 3 to 6 miles per hour breeze, typical of those measured on the surface of Venus, would have the force of waves in a terrestrial gale.

Lest some still find the planet romantic, there is still worse to come. To the above bad news must be added the composition of the atmosphere, which is 95 percent carbon dioxide. Floating in that atmosphere, above the surface, we find clouds of rather concentrated sulphuric acid, with a concentration greater than that in a standard car battery (which as we know can cause serious acid burns if it comes into contact with skin). To add to the malefic mix, there is even a small quantity of fluorosulphuric acid, and, as the British meteorology expert Garry Hunt once put it, "With that delightful substance, one can even dissolve rocks!" Venus thus has a serious acid rain problem too, besides being a toxic inferno.

The Venera probes also took color images of the planet. They showed that the light on the surface of Venus is an infernal orange-red color: as the sunlight filters through the thick layers of cloud, only the red part is capable of reaching the surface. Even so, Soviet scientists reported that the light level on the surface of Venus was "the same as on a cloudy day in Moscow."

No astronaut standing on the surface of Venus would "appreciate" the view for long. Unless he or she were protected by a capsule and a spacesuit far beyond anything that the Apollo astronauts ever took into space with them, the astronaut would be roasted, squashed, asphyxiated, poisoned, burned, and blown away, all simultaneously and terminally. Far from being the planet of love, Venus is the closest that one can come to the medieval images of hell.

Could this torrid Venus have been the serene Star of Bethlehem? The truth is that this is an old saw, and was an old saw already in the nineteenth century. Writing in *The Scenery of the Heavens* in 1890, the English amateur astronomer J. E. Gore observed the following:

> When it [Venus] happens to be a morning star about Christmas time (this occurred in 1887, and again in 1889) it has often been mistaken by the public for a return of the "Star of Bethlehem"! Whatever the star of the Magi was, one thing is certain, it was not Venus. It seems indeed absurd to suppose that the "wise men" of the east should have mistaken a familiar object, like the planet Venus, for a strange apparition. That it was familiar to the ancients we know. . . . Indeed, it seems impossible that so conspicuous an object should remain unnoticed for any length of time.

As we have seen, the Magi were part astrologers and part astronomers, and, as such, they would have known the five main planets and their movements in the sky very well indeed. In fact, it would have been their prime function as astrologers to have this information and to know how to interpret it correctly. Venus was also well known to the Chinese, the Egyptians, and the Babylonians, as it was to the Romans, the Greeks, and probably to any civilization that has existed on Earth since the very first cave dwellers, and it is unlikely anyone would still be fooled by Venus.

Venus is visible in the sky almost continuously, except for a few days every year and a half, when it passes between the Sun and the Earth, and for a few weeks more, nine months later, when it passes behind the Sun. Only rarely is Venus so close to the Sun in the sky that it is not visible as either an evening or a morning object. It is never invisible for more than two or three weeks and, what's more, it is occasionally—even at conjunction when it passes close to the sun in the sky—visible briefly both before sunrise and after sunset on the same day. Therefore, if we assume that the sight of Venus triggered the journey of the Magi to the Holy Land, we also have to ask just *why* they went when they did and not in another year, or even in another month, because Venus is visible virtually every month of every year. Patrick Moore, the British broadcaster and popularizer of astronomy, has put it more strongly and succinctly: "If the 'wise men' were fooled by Venus, they could not have been very wise!"—a single phrase that sums up the situation beautifully.

Venus on its own, then, was not the Star of Bethlehem, and the suggestion that it might have been is quite absurd when we consider it critically. It was simply not unusual enough to have moved the Magi who had been waiting for a sign for so long. They needed a far more convincing cue. What is not impossible, though, is that the Star of Bethlehem might have been Venus *in combination with* another body.

Was It Halley's Comet?

For some, a visit to a small Italian town offers a clue to the identification of the Star. There is a chapel in the northeastern town of Padua, just twenty or so miles west of Venice, the Scrovegni Chapel. As you climb the stairs and enter it, you soon spot a fresco above you on the second tier from the top. For some, this depiction is documentary evidence for their theory of the Star.

The Florentine painter Giotto di Bondone (ca. 1267–1337) painted this fresco sometime between 1302 and 1304. The exact date of the painting is not recorded, but we know that the chapel, newly built, was dedicated on March 25, 1303, and consecrated on March 25, 1305. The land for the building had been acquired on February 6, 1300, and it is assumed that Giotto was commissioned to paint the frescos for the chapel in 1301. What the artist originally had in mind for the subject matter of the frescos is unclear. But we do know that he was powerfully influenced by a particular astronomical event. What he saw, which so impressed him, was Halley's Comet, although it was not so named until the eighteenth century.

The comet appeared in 1301 and was apparently first seen by European observers on September 1. We now know that this was nearly two months before perihelion, when the comet comes closest to the Sun. This is an unexpectedly early moment to see Halley's Comet for the first time on its

Figure 4.2. *Adoration of the Magi* by Giotto di Bondone (1266–1336), at Scrovegni Chapel, Padua, Italy. (Cameraphoto/Art Resource, NY.)

return, suggesting that it was unusually bright at the time. Giotto almost certainly saw the comet at this apparition. He was reputedly stunned by its beauty—and it provided the subject matter for one of the frescos.

This particular fresco is entitled the *Adoration of the Magi* (fig. 4.2). It depicts the Nativity with the three kings paying homage to the infant Jesus. In the sky above the stable, you clearly see a comet with a flowing tail behind it. In fact, this was the first time that anyone had faithfully and accurately reproduced the appearance of a comet in a painting or drawing. Prior to this, all cometary images that we know of had been highly stylized and held minimal scientific information. In fact, several centuries were still to pass before the majority of artists and scientists regularly attained this eye for detail shown by Giotto. Seeing this comet must have been a particularly striking and memorable experience for him. Through this fresco Giotto clearly suggested that a comet like the one he had recently observed was spectacular enough to have been the Star of Bethlehem. What Giotto di Bondone did not know at the time, and could not ever have known, is that, several centuries later, this same comet *would* be suggested seriously as having been the Star of Bethlehem.

Was Giotto's comet really the one we know of today as Halley's Comet? Most authorities say yes—the comet painted in the Scrovegni fresco is genuinely Halley. For this reason, when the European Space Agency (ESA) launched its brilliantly successful probe to the comet in 1985, it was named Giotto, and the fresco became an unofficial symbol of the mission. Still, lately some authorities have expressed doubt that this is truly Halley's Comet; they have found the evidence wanting. They suggest the comet painted by Giotto was, in fact, a later one observed in 1304. Among the skeptics is the well-known British cometary expert, astro-historian, and expert on the Star of Bethlehem, David Hughes.

Gary Kronk, the American amateur astronomer and comet expert offers a clue. He has compiled information for hundreds of the brightest comets seen in recorded history. In his seminal work on all the important comets observed over the centuries, *Comets—A Descriptive Catalog,* Kronk does not even mention the 1304 comet, indicating that he does not consider it to have been a particularly bright and important one. But other research uncovers different findings. For example, records of both Comet Halley, in 1301, and the comet of 1304 are found in Chinese, Korean, and Japanese chronicles. A comparison of these seems to confirm the impression that, although the 1304 comet was observed for a longer period of time than Halley's Comet had been three years earlier (that is, seventy-four days versus exactly two months), it was Halley's which was considered to be the more spectacular of the two.

More evidence for this conclusion comes from revisiting the Scrovegni Chapel. Giotto's fresco of the Nativity is on the second tier from the top of the chapel. This means that the comet was probably one of the very first parts of the walls to be painted, that is, after Giotto had seen Halley's Comet but well before the 1304 comet appeared in the heavens.

At present, then, there is no conclusive reason to suggest that the Star in Giotto's *Adoration of the Magi* is anything other than Comet Halley. The same applies to the modern suggestion that Giotto had astrological and not theological motives for his representation of the comet in the scene. Giotto's reputation for combining naturalistic observations (that is, painting what he had seen) with theological tradition is well founded and there is no doubt that he painted the comet as a serious and considered religious and scientific statement. The Star is painted as a comet because that is what many theologians of the day believed it was. Many of the experts who have studied the

Star of Bethlehem still hold this belief, nearly seven centuries after Giotto's time.

After Venus, Halley's Comet is the object most frequently suggested to have been the Star of Bethlehem. One of the reasons why it is such a popular explanation may be because it is the most famous comet of all and the one best known to the general public. It is neither the brightest nor the most spectacular comet, but it is the one that comes around regularly. Occasionally, events are even described in the press or in books as being "as regular as Halley's Comet."

Only a minority of comets return regularly like Halley's Comet. A dividing line is drawn between comets that have an orbit period of two hundred years or less, and those with longer orbits. The former are called "periodic comets"; the latter are collectively called "nonperiodic" or, more correctly, "long-period comets." As of 1995, 170 periodic comets, compared to 671 long-period ones, were known. Of the periodic comets, most have periods of less than twelve years, and only a small fraction have orbits as long as or longer than Halley's. Halley's Comet is thus rather unusual in this sense, even apart from being much brighter than any other periodic comet.

After its last appearance in 1986, people's general impression of this famous visitor was that it was terribly disappointing and uninspiring, and they commonly do not regard it as one of the brightest and most spectacular comets of the last two decades (which indeed it was). Their experience was probably colored by the fact that Halley's Comet was comparatively faint in 1986—in fact, it was the faintest sighting on record, in over two thousand years. It is no consolation to hear that this was due to bad luck, that the comet and the Earth were just in the wrong place at the wrong time. If Halley's Comet had arrived just a few days later (a two-week delay would have been sufficient), it would have passed very close to Earth and become extremely brilliant. Indeed, of the twenty-eight occasions on which it has been recorded since 240 B.C., Halley's Comet has failed to get brighter than magnitude 2 on only two occasions—those in 163 B.C. and in 1986. On a number of occasions, about a quarter of all recorded appearances, it has reached negative magnitude. In the year A.D. 837 the comet passed extremely close to Earth and would have even been as bright as Venus for a few days.

During its long observational history, probably dating back to the winter of 1059 to 1058 B.C., Halley's Comet has been seen on many historically famous and important occasions. Perhaps the most famous of these was in 1066. In that distant year, Harold Godwinson of England "reneged" on an agreement to allow the throne of England to pass to the Normans on the death of Edward the Confessor. Harold seized the throne and had himself crowned king. Why did Harold take this extreme step? The simple reason is that Edward the Confessor had also promised him the throne of England, supposedly saying on his deathbed, "Into Harold's hands I commit my kingdom," a singularly evenhanded approach by Edward to the two rivals. News of this action reached William of Normandy, who naturally felt that he was the rightful heir to the throne. William, angered by this "betrayal," resolved to gain the throne that was rightfully his by conquest on the battlefield. In the meantime, during the summer of 1066, while William prepared his invasion of

England and his revenge on Harold, Comet Halley appeared. On this occasion it became quite bright and extremely prominent in the sky because it passed rather close to Earth. A month after perihelion it went past at a distance of just 10 million miles. This is still one of the closest known approaches of a bright comet to our planet and is the closest approach that Comet Halley has made to the Earth since A.D. 837.

As the Normans and the English were preparing for war, Halley's comet was observed in China, Japan, Korea, and Europe for two months, from April 3 until June 7, 1066. During this time its magnitude probably reached about −2, although some evidence suggests that it may actually have been even brighter. Chinese observations speak of a tail more than 20 degrees long. An Italian record states that "it looked like an eclipsed moon, its tail rose like smoke halfway to the zenith." Back in England, Harold and his subjects saw this spectacular comet in the sky and took it (correctly) as a sign of an impending catastrophe. Harold received his comeuppance shortly afterwards. Having successfully beaten off a Viking raid on the north of England, at the Battle of Stamford Bridge, his army had to return south, by forced march, on receiving the news of the Norman invasion on the south coast. It must be said that Harold Godwinson, the fleeting king of England, was singularly unfortunate in many respects. Not only did the Viking raid occur precisely at the moment when William was preparing his invasion, but it was even led by his own stepbrother, Tostig, in the company of the Viking king of Norway, Harald Hardrada.

Thanks in part to poor English battle tactics and to a new war technology, William won the ensuing battle. This technology was unexpected: it was not a new weapon as such, not a new bow. It was stirrups. These allowed horsemen to lead their horses to the left or right, changing their direction at will. Harold and the Saxons in general did not use horse cavalry; in fact, their only use for horses was for transport. One of the reasons why cavalry (or elephant) charges were not frequently used in battle by armies prior to this was because they were virtually impossible to control without stirrups: the cavalry would head in a more or less straight line, at great pace. If the infantry facing the charge opened its lines, the cavalry could pass straight through the gap, doing little or no damage, totally unable to stop or to change direction. The Norman introduction of stirrups allowed them a control of cavalry which was devastating in the field of battle. Initially, Harold's army held the high ground, just outside the Sussex town now called Battle, and their shield wall, bristling with spears, proved totally impervious to the Norman infantry, while the Norman cavalry was useless and could have been devastated by a rain of arrows from Harold's archers. By feigning withdrawal, the Normans induced the Saxon army to break the shield wall and charge down the hill in pursuit of the supposed rout. The Norman cavalry could now turn on the astonished Saxons who, without the advantage of the high ground and mutual defense, were literally hacked to bits. By the time Harold's army had realized their mistake and reorganized, the only option left to them was a desperate last stand. The carnage must have been terrible, for the Normans would eventually call the site of the battle Senlac Hill, from two Norman words meaning "lake of blood." Harold was killed, and as a result, his rival's name changed, for the history books, from William the Bastard to William the Conqueror.

Figure 4.3. Comet Halley in the Bayeux Tapestry.
(Giraudon/Art Resource, NY.)

The appearance of the comet, showing King Harold and his courtiers pointing at it, with Harold visibly trembling on the throne in terror, is recorded for posterity in the Bayeux Tapestry, supposedly embroidered by William the Conqueror's wife (fig. 4.3).

In 1456, during another rather bright appearance of the comet, legend has it that Pope Calixtus III excommunicated Comet Halley as an agent of the devil. This story is somewhat controversial, and some scholars believe it is an apocryphal event and merely the result of later anti-Catholic propaganda, designed to ridicule Catholics and their practices.

But to follow the Comet Halley connection with the

Star of Bethlehem, we must now jump four centuries forward in time, from Giotto's painting of the Nativity in the fourteenth century to the start of the eighteenth—and, incidentally, to the naming of the comet. In 1705, the English astronomer Edmond Halley published a six-page pamphlet on his determinations of cometary orbits. The meat of the work was a table of twenty-four cometary orbits. Halley noticed that three of the comets that were listed, those observed in 1531, 1607, and in 1682, had extremely similar orbits. He also noticed that the comet appeared to return every seventy-five and a half years. Starting from this information, Halley arrived at the correct conclusion: that the three comets were really one and the same, and that this object returned in a regular cycle. From this calculation he predicted that the same comet would return, once more, in 1758. Knowing that he would never live to see the prediction fulfilled, Halley made this famous statement before dying: "Wherefore if according to what we have already said it should return again about the year 1758, candid posterity will not refuse to acknowledge that this was first discovered by an Englishman." When he died, in 1742, at the advanced age of eighty-seven, the comet's return was still many years in the future. His prediction was correctly fulfilled, but only just, for the comet was finally observed late in the year, on Christmas Day 1758. To the enormous disgust of the French scientific community, the first sighting was not accomplished by the great French comet hunter, Charles Messier, who had been dubbed "the Ferret of Comets" by the French king because of his extraordinary ability to locate them. Messier had already spent many months searching for the comet, but to no avail. It was a farmer named Johann Palitzsch, living near Dresden in what was then Saxony and is now eastern Germany, who claimed the honor of the first sighting. Messier had to be satisfied with a later, independent discovery of the comet, in January 1759, and was reportedly none too pleased to be upstaged. Halley had, in fact, hedged his bets after making his first prediction, having correctly guessed that perturbations to the comet's orbit, caused by the pull of gravity of Jupiter, might delay the comet. He revised his initial prediction to read "the end of 1758, or the beginning of the next [year]."

Messier was, understandably, bitterly disappointed that his months of effort had come to nothing. The French scientific community, in general, which had dedicated a great deal of effort to verifying Halley's prediction, was most unhappy that a Saxon amateur astronomer should beat their finest mathematicians and astronomers to the recovery of the comet. There was even some measure of bad feeling between the countries, adding to an already prevailing current of nationalist feelings. The comet was named after Halley, the Englishman, and the French astronomer Messier was left to "ferret out" other comets.

Let us now look closely at the possible connection between Comet Halley and the Nativity. On this 1758–59 sighting, the actual interval between successive perihelia (the period of the comet) was rather longer than previously—seventy-six and a half years—because the comet did not reach perihelion until March 1759. This, too, was suspected by Halley. Calculating backwards, it seemed possible that Comet Halley might have been visible at the time of the Nativity. If its period were 76.5 years, twenty-three revolutions would have taken 1,759.5 years to complete. The simple sum, 1759 − 1759.5, suggested that the comet might have been visible at exactly the right time, around the year 0. This led astronomers to make the connection of Halley's Comet with the Star of Bethlehem, no doubt inspired by the very bright apparition of the comet in 1759, one of the brightest in the last thousand years.

Although Halley's Comet is visible from the extreme south of Europe only when brightest (as in 1986, when it passed far into the southern sky), the combination of its brightness and the coincidence of date made this comet an obvious candidate to be the Star of Bethlehem. The fact that it was, by then, the most famous comet of all and the subject of numerous stories and anecdotes, simply made the connection seem even more obvious to those who suggested it—especially those who might have visited the Scrovegni Chapel.

Halley's Comet, the Star of Bethlehem? It seemed almost too beautiful to be true. Alas, it *was* too beautiful to be true. The orbit of Halley's Comet was investigated a little further when it appeared again in 1835, and it became evident that its period was not 76 years (as is popularly believed), nor 76.5 years (the interval between its apparitions in 1682 and 1759), but almost exactly 77 years on average. The exact interval between the comet's returns depends on the slight changes to its orbit caused by the gravitational pull of Jupiter and Saturn. The interval has thus varied slightly over the centuries, from 74.5 years to 79.5, depending on whether Jupiter and Saturn advance or delay it in its orbit. Such is the push and pull that exists within our solar system.

In the middle of the last century, English amateur

astronomer John Hind investigated the orbit of Halley's Comet and suggested that a comet seen in 11 B.C. was actually Comet Halley. Hind correctly identified many old appearances of the comet but, in this case and in five others, he was in error. This misidentification is even more puzzling because the best present-day catalogs of ancient Chinese observations of comets list objects in 10 B.C. and in 12 B.C., but none in 11 B.C. Even the 10 B.C. comet seems to have been an error in the ancient chronicles and is now often referred to as the 10 B.C. "ghost event."

Comet Halley was the object observed in China during an observation from August 26 to October 20, 12 B.C., an occasion on which the comet was reasonably but not outstandingly bright. It passed from Canis Minor (the Little Dog) in the morning sky to Scorpius in the evening sky before disappearing behind the Sun. Although there seem to be few important details in the Chinese accounts of the comet, the Romans suggested that its appearance marked the death of the Roman statesman and general, Marcus Agrippa. As we saw in chapter 1, few important citizens and no emperor ever died in Rome without some portent being attached to the event, often a celestial one. A typical interpretation would be that a comet was a sign that a dead emperor had become a god, and the comet was the manifestation of his soul being lifted into the heavens.

The fact that Comet Halley was visible in 12 B.C. makes it a most unlikely candidate to be the Star of Bethlehem. Although the exact date of Christ's birth, as we have seen elsewhere in this book, is not exactly known, it was almost certainly between 7 B.C. and 4 B.C. Alternative estimates of the date that fall outside this range delay the Nativity even further. Even taking the earliest possible "reasonable" date for the Nativity—7 B.C.—the delay of nearly five years between the appearance of Halley's Comet and the birth of Jesus makes the association with the Star of Bethlehem particularly unlikely. If the Magi really took nearly five years to make their journey, they were evidently in no great hurry.

With some regret, and with some historical suggestions and one of the great Giotto frescos to the contrary, we must conclude that Comet Halley was not the Star of Bethlehem.

Was It a Conjunction of the Planets?

Various popular theories to explain the Star of Bethlehem have been based around an astronomical phenomenon called a "conjunction" of planets. A conjunction is when two or more planets line up in the sky in such a way that they appear close together, yet, in reality, they remain millions of miles apart. Occasionally, the alignment is so perfect that one planet actually passes in front of another, in what is termed an "occultation." Conjunctions have always been particularly important in astrology, and it requires little effort of imagination to understand that the Magi, as astrologers, would have thought long and hard about the implications of the different conjunctions they saw in the ancient sky.

We will now examine two conjunction theories—one in some detail, the other quite briefly.

The Conjunction of Jupiter and Venus in 2 B.C.

In the December 1968 edition of the magazine *Sky and Telescope,* Roger Sinnott, now an associate editor of the magazine, wrote a short but seminal article about the Star. It is regarded as one of the most important ones published on this subject in the past forty years, since he attacked the problem of explaining the star on a highly scientific basis.

Sinnott started from the hypothesis that the "star" might actually have been a conjunction of two or more planets. He then embarked on a massive and, in a pre-computer age, highly tedious study to identify the best candidate or candidates. This study was done by hand, from tables of planetary positions ranging in date over two millennia. His study was made even more difficult because of the contradiction between the different clues in Luke's Gospel as to the date of the Nativity. Because these contradictions had not been satisfactorily resolved, Sinnott was obliged to search an interval of almost twenty years, from 12 B.C. through to A.D. 7, for a possible explanation of the Star of Bethlehem.

Sinnott found around two hundred conjunctions of at least two planets, and twenty multiple groupings of at least three planets. In 90 percent of these, the planets were either too faint to be striking or too close to the Sun to be visible in the sky. Studying the candidates one by one, he narrowed them down to twenty, each of which he then studied in more detail.

Of the twenty candidates, some would not have been visible in the Near East and obviously had to be discarded. Only four multiple groupings, none of them particularly

TABLE 4.1 The Most Spectacular Conjunctions of Two Planets Observable between 10 B.C. and A.D. 1

Date	*Planets*	*Local Time*	*Separation*
August 12, 3 B.C.	Venus + Jupiter	03:44–05:23	12 minutes
June 17, 2 B.C.	Venus + Jupiter	19:04–22:02	3 minutes

spectacular because all occurred low on the horizon in a bright twilight sky, and six ordinary conjunctions remained. Most of the ordinary conjunctions were also comparatively unspectacular, but two bucked the trend, as shown in table 4.1. Both of these conjunctions occurred in the constellation of Leo. The first occurred rather close to the Sun (21 degrees away) and would have been seen low in the bright morning twilight sky. The second would have been very different: the two planets would still have been comparatively high in the evening sky above the western horizon after dark. The conjunction was such that the two planets would not have been separable with the naked eye at the time both of them set—they appeared as one.

This conjunction, it turns out, was even more spectacular than Sinnott calculated because, as seen from Jerusalem, the two planets almost touched: the gibbous Venus, with 80 percent of its disk illuminated, passed just south of the slightly larger disk of Jupiter. When the two planets were closest, they would have been 14 degrees high in the west, in a completely dark sky. From anywhere in the region the conjunction would have been tremendously spectacular and obvious to almost anyone who looked at the sky at dusk, between sunset and the moment when the two planets were closest together and seemed to fuse into one. A later study has shown that, a few hours later, a partial occultation even took place, with the disk of Venus partially covering Jupiter.

A Conjunction of Jupiter, Saturn, and Neptune

Others have made similar claims. One of the most remarkable suggestions was published some ten years ago in a reputable publication. In it, the author suggests that the Star of Bethlehem could have been "a special conjunction of the planets Jupiter, Saturn and Neptune." This statement is truer than it might seem. Neptune is totally invisible without a telescope (whose invention would come just 1,600 years too late to be of service in this case), and the planet was not discovered until 1846. The first planetary conjunction involving Neptune to be observed was in the winter of 1613–14 by Galileo Galilei. He was observing Jupiter and, in one of his sketches, recorded a faint star close to the planet—one that was not there in his next observation. Galileo never realized that his "faint star" was actually a new planet, and the observation lay, unrecognized, for more than three and a half centuries, until the 1980s.

But, to have been visible to the Wise Men, this triple conjunction needed to be more than special. In fact, Neptune was not even close to Jupiter and Saturn in the sky around 5 B.C., which would have made the conjunction (implying close proximity in the sky) totally impossible. It may be of interest that triple conjunctions between Jupiter and Neptune occurred in 25–24 B.C. and then between Saturn and Neptune in 22–21 B.C. At no point, however, were the three planets ever close together in the sky between 25 B.C. and A.D. 10. The last conjunction of any kind visible during this interval was between Jupiter and Neptune in 12 B.C.

The Conjunction of Uranus and Saturn

If the Star of Bethlehem was not Venus, nor a conjunction of planets as described above, could it have been the conjunction of other planets? There is a new theory attributed to John Harris and described on an internet site maintained by the British Methodist minister and amateur astronomer, the Reverend Phillip Greetham. This theory is alluded to in a number of Star of Bethlehem Websites, which have become a minor industry. The idea is this: the Magi could have observed the planet Uranus when it passed close to Saturn in 9 B.C. and later recovered it when it passed close to Venus in April 6 B.C. Being a new planet for the Magi, it would have been regarded as a great sign to them.

It turns out that Saturn and Uranus did have a conjunction around this time, although I cannot confirm all the details given by the Reverend Greetham. Between 10 B.C. and A.D. 1 there was just one conjunction of Uranus and Saturn, although there were no fewer than ten conjunctions of Uranus and the faster-moving Venus. The Uranus-Saturn conjunction occurred on February 5, 9 B.C., with the two planets separated by 66 arcminutes, slightly more than one degree. (Curiously, Uranus was in Pisces at the time, and Saturn was in Aquarius.)[1] But this conjunction would hardly have been a spectacular event because it occurred very close to the Sun. Even though the two planets would have been fairly close together for several weeks, both would have been very low in the twilight sky. For this reason, I have grave doubts that Uranus would have been noticeable at any time when it was close enough to Saturn for the planet to have drawn attention to it. A further problem with this theory is that there is another object, unknown at the time, which can get just as bright as Uranus. This other heavenly body is Vesta, the brightest of all the asteroids and one of the largest of this swarm of hundreds of thousands of rocks that orbit the Sun between Mars and Jupiter. It should also have been detected by the Magi, who could have mistaken it for a new planet. If the Magi were capable of "discovering" Uranus, they should certainly have seen Vesta, too.

Was It an Occultation by the Moon?

Occasionally, during the course of its monthly orbit around the Earth, the Moon passes in front of a star or a planet, hiding it for a time. This phenomenon, known as an occultation, can be quite spectacular. If a bright star, or planet, suddenly reappears at the border of the Moon's disk, particularly the border which is in shadow, the event is really quite impressive. A bright planet emerging from behind the Moon's darkened border is a stunning sight. Could the Star of Bethlehem have been an occultation of a bright planet? Some observers have, in the last few years, made this suggestion. One version was published in a reputable professional publication, so it has to be taken very seriously.

Occultations of planets by the Moon are not rare. In fact, when I used the computer program "Dance of the Planets," which calculates the position of the stars and the planets accurately for several thousand years into the past or the future, to conduct a search, I found no less than 304 occultations of planets by the Moon just between 20 B.C. and A.D. 1. Of course, the Magi would have missed a lot of them: many involved faint, outer planets or Mercury, and would have been difficult or impossible to see with the naked eye. When these are excluded and we look only for occultations of the four brightest planets (Venus, Mars, Jupiter, and Saturn), this number is reduced to "only" 170 occultations during the twenty-year period.

Like eclipses of the Moon or Sun, not all occultations are visible from all points on the Earth. This is due to several reasons. Sometimes the Moon is simply not above the horizon and is thus invisible from our viewing point. Sometimes the occultation happens during the day and is thus invisible without a telescope. And sometimes there is the parallax effect. Because the Moon is relatively close to us, astronomically speaking, it is displaced in the sky depending on one's exact viewing point. Although the Moon might cover a planet as seen from some regions of the Earth, it may miss the planet completely from other vantage points. This parallax effect can shift the moon as much as three times its own diameter in the sky, depending on whether it is observed from one extreme of our planet or from the other. So our task is to determine how many of these 170 occultations that occurred between 20 B.C. and A.D. 1 could have been seen by the Magi.

If we exclude all occultations that were not visible from Babylon, the number of events drops to just 17—about one per year. However, not all of these would have been visible to the Magi: most occurred during the day and would have been invisible without a telescope (which the Magi most certainly did not have). So our possibilities are narrowing.

In fact, only six occultations of bright planets occurred

between 20 B.C. and A.D. 1 that were visible from Babylon and happened at a time when the Sun was below the horizon. But we have to eliminate one more: here, the Sun was about to rise and the occultation would have been quite invisible because the Moon was so low in the dawn sky.

Let's consider the remaining five possible candidates: two occultations of Mars, two of Saturn, and one of Jupiter. Of these five visible-in-Babylon occultations, four occurred close to Full Moon and would have been of limited impact because of the enormous lunar glare. The most striking would have been the occultation of Jupiter, observed on July 13, 17 B.C. As the thin crescent Moon rose above the horizon, slightly before twenty past one in the morning local time in Babylon, it would have been accompanied by a brilliant star. The Moon and the star were between the Horns of the Bull in the constellation of Taurus. This star, the planet Jupiter, would have creeped gradually closer to the crescent over the next two hours. At 3:31 A.M., almost an hour and a half before sunrise, with the sky still completely dark, Jupiter would have passed behind the Moon, disappearing over the course of a minute or so. Around 4:05 A.M., the planet would have reappeared, near the top (the north) of the Moon, against the darkened part of the disk. Sunrise would occur slightly more than 50 minutes later, and, with Jupiter more than 30 degrees above the eastern horizon, the sky near the Moon would still have been dark. The reappearance of the planet would have been stunningly beautiful as it was reborn in the dawn sky.

Could this have been the Star of Bethlehem? The answer once again is most obviously "no," as this phenomenon occurred long before the date of the Nativity. Even assuming that our date was seriously in error, there is a further problem with this theory. During the first century B.C., no fewer than 386 occultations of either Venus or Jupiter by the Moon could potentially have been observed. Even when we eliminate all those that were not visible in a dark sky from Babylon, the Magi could still have seen eight really spectacular occultations (five of Jupiter and three of Venus) in just one century—hardly the mark of the rare event that the Magi had been seeking for so many years.

In fact, the two most spectacular occultations of the entire first century B.C. occurred within eighteen months of each other, in the autumn of 45 B.C. and the winter of 46 B.C. During the early morning of March 4, 46 B.C., the Magi would have witnessed a spectacular occultation of Jupiter in the constellation of Ophiuchus near the Moon, which had just passed First Quarter.[2] This occultation would have lasted from 4:36 to 5:25 A.M. local time. Even more spectacular would have been the occultation of Venus seen on November 18, 45 B.C. The slim, waning crescent Moon, in the constellation of Virgo, would have occulted Venus, low in the predawn sky, from 4:25 to 5:33 A.M. local time. If any occultation were to attract the interest of the Magi, the combination of these two, separated by only a year and a half, would surely have been a sign for them.

Once again, we know (or we can at least safely assume) that the Magi did not go to Jerusalem in 45 B.C., as surely they must have done if an occultation was the Star they were awaiting. See table 4.2 for a summary of lunar occultations that occurred from 10 B.C. to 1 B.C., the dates closest to the Nativity, with the five major planets (Ura-

TABLE 4.2 Summary of the Number of Occultations of the Five Brightest Planets Occurring between 10 B.C. and 1 B.C.

	Mercury	*Venus*	*Mars*	*Jupiter*	*Saturn*
Total number of occultations	24	20	28	16	14
Number of occultations visible from Babylon	1	2	3	2	1
Number of occultations visible in the night sky from Babylon	0	0	2	0	0

NOTE: We see that there were, for example, no less than sixteen occultations of Jupiter by the Moon in this ten-year span. Only two out of more than a hundred occultations in total would genuinely have been visible to an observer in Babylon.

nus, Neptune, and Pluto, which are invisible to the naked eye, or just barely visible, are excluded).[3]

Many more spectacular occultations would have been visible during the preceding centuries and, once again, one is able to select the "right one" only by knowing the date of the Nativity in advance. The Magi did not have this information and would not easily have known how to select one special occultation over many others and would likely choose none. But let's take a look at a theory that suggests how the Magi *might* have known which was the correct occultation.

Occultations and the Coin of Antioch

Recently there has been a lot of interest in a particular occultation of Jupiter that was seen in the constellation of Aries in 6 B.C. This occultation is intriguing because it may be referred to indirectly in an ancient coin found in

Antioch, a city on the Orontes River in southern Turkey that was once the capital of ancient Syria, well to the north of Jerusalem. In fact, there are really two occultations of Jupiter by the Moon that have been cited as possible explanations of the Star of Bethlehem. On March 20 and April 17, 6 B.C., the Moon occulted Jupiter in the constellation of Aries. These were the central two of a series of four monthly occultations of Jupiter by the Moon. They are the only occultations of Jupiter that occurred between the beginning of 7 B.C. and the end of 4 B.C.

The leading exponent of this theory is Michael Molnar, an American astronomer at Rutgers University in New Jersey. Also an expert in ancient coins, he was struck by one particular coin issued in A.D. 13 or 14 and by a very similar one from about A.D. 55. Both coins show a ram looking at a bright star close to a crescent Moon. The one from A.D. 55 is now referred to as the "Antioch coin" (fig. 4.4). Molnar comments that Aries, which is the constellation of the Ram, astrologically rules over Palestine and Judea as well as various other states and thus would have been closely associated with the Jews. He suggests that the coin of A.D. represents an occultation of Venus by the Moon, and the one from A.D. 13–14 a conjunction of Mercury and Jupiter in Aries in A.D. 7. Molnar's theory is that, by extension of what is known of the astrology of the period, a significant event for the Magi would have been an occultation of the planet Jupiter by the Moon in the constellation of Aries.

Such an occultation would signify to the Magi the birth of a king, while the constellation in which it occurred (Aries) would tell them that the occultation fore-

Figure 4.4. The Antioch coin. (From Andrew M. Burnett, Michel Amandry, and Pere Pau Ripoll, *Roman Provincial Coinage*. London: British Museum Press; Paris: Bibliothèque Nationale, 1992.)

told the birth of a king in Judea. Investigating further, he found two lunar occultations of Jupiter in Aries that occurred at just about the right moment to be the Star of Bethlehem. The two occultations were the ones of 6 B.C. discussed above. This theory has created a great deal of interest and has been enthusiastically received by various astronomers and science writers, including my ex-colleague, Marcus Chown, at the Physics Department of Queen Mary and Westfield College (London University), who wrote an article about the theory for the December 1997 issue of *New Scientist*.

However, as we can see from table 4.3, the circumstances were not particularly favorable for observing these occultations. Although the March and April occultations were theoretically visible from Jerusalem, we must also consider practicality. The April 17 occultation occurred near the heliacal rising of Jupiter, with the Moon exactly one day

TABLE 4.3 Details of the Four Occultations of Jupiter by the Moon Observable during 6 B.C.

Date	*Time*	*Moon's Phase*	*Visible from*
February 20	17:56	+6%	—
March 20	14:06	+1%	Jerusalem
April 17	09:30	−1%	Babylon, Jerusalem
May 15	04:30	−8%	—

NOTE: The first and last events were invisible from Babylon and Palestine because the Moon was below the horizon when the occultation occurred.

before New, when it is unlikely to be visible. In principle, the Moon can only just be seen, in perfect conditions, sixteen hours from New but, in practice, this is almost impossible except in extremely rare circumstances. In fact, the thin crescent Moon is difficult to see with the naked eye less than a day and a half from New Moon. Furthermore, the occultation occurred around midday in Jerusalem and Babylon, and a thin crescent Moon would be totally invisible in the sky so close to the Sun during daylight. In any case, Jupiter simply cannot be seen in daylight except under truly exceptional conditions by people with unusually good eyesight. This occultation would thus have been totally invisible.

The March 17 occultation occurred after sunset. It was not visible in Babylon because Jupiter had already set there at the time, but it was visible in Jerusalem, if only marginally. The Moon was a thin crescent, fourteen hours old, well below the accepted minimum for visibility. It set at 6:24 P.M., just 35 minutes after the Sun. At the time of the occultation, 5:58 P.M. local time in Jerusalem, the Sun was 3 degrees below the horizon and the Moon was 5 degrees above it, less than half an hour before it set.

Thus it is extremely unlikely that the March occultation could have been observed, and the April occultation would have been impossible to observe without a telescope. Molnar points out these difficulties, although he comments that Jupiter can be seen with the naked eye as close as 12 degrees from the Sun. If so, the planet could have been visible to the Magi on both occasions even if the Moon were not. More controversial is his suggestion that it was not even necessary to *see* the occultation for it to have been a powerful sign for the Magi. He suggests they could have calculated when it would occur and realized its significance even if they were then unable to observe the phenomenon themselves. King Herod, on the other hand, even if he were aware of the occultation, which is most unlikely, would have ascribed no particular significance to it. The timing of the occultation is also convenient, as we have seen, because the biblical evidence suggests that Jesus was born around the month of March or April of either 6 B.C. or 5 B.C., precisely when the occultations would have occurred.

None of the four occultations of Jupiter in the series of events referred to above could have been observed from Babylon, however. If we assume that the Magi began their trip from there, the occultations could not have been seen by them. Had the Magi come from Persia, the same situation would have been true: only one of the occultations occurred above the horizon in Tehran, and that was in daylight during the afternoon. In other words, unless the Magi were in Jerusalem and they had extremely acute eyesight for the March 6 B.C. occultation, which was low in the bright twilight just after sunset, these occultations could not have been the visible sign for which they were waiting so anxiously.

For Molnar's theory to be tenable, we must assume that the Magi were able to interpret correctly an event they could not have seen. However, of all the theories that have been proposed at different times, this is still one of the most credible, since the Magi could have *suspected* that the occultation had occurred. If so, it would have been a strong indication to them that something was going on in Judea, but they would still have missed actually seeing the final and definitive sign that would announce the birth of the new king.

In this chapter we have traveled some distance across time and place in our search for the Star—from the recent space probes and landings on bright and infernal Venus to Giotto's fresco of the Nativity and Halley's Comet in a nearly seven-hundred-year-old Italian chapel. Rejecting Venus and the comet as possible Stars, others have brought us data and speculation on conjunctions and occultations, both seen and apparently unseen. Could the Magi have understood such complex celestial occurrences? Did they know their significance? Or were they looking for a more simple, clear-cut sign, a literal shout from the heavens? We'll look for the answers in the next chapter.

5

Shooting Stars and Fiery Rains

Over the years, various publications have suggested a radically different theory to explain the Star of Bethlehem: was it a meteor rather than a star, planet, or combination of planets?

Most people have seen a shooting star, more correctly called a meteor. If you have stood outside on a warm summer's night in August, anywhere in the Northern Hemisphere, you are almost certain to have seen bright, typically yellow meteors coming from the northeast. These meteors are the Perseids, named for the constellation of Perseus, from which they appear to radiate. In 1838, the Yale University librarian, Edward C. Herrick, pointed out that the Perseids appeared every year around August 9 and 10. Herrick also popularized the name the "Tears of Saint Lawrence" for this shower. He pointed out that Irish peasants referred to these meteors as the burning tears of Saint Lawrence, which was appropriate because Saint Lawrence was martyred by being cooked alive over red-hot coals—after being whipped with scorpions—during one of the early crackdowns on the Christian church. According to legend, just before he died he called out to his torturers, "Turn me over, I'm done on this side!" The tears of fire falling from the sky on his feast day, which is celebrated on August 10, are an appropriate memorial. The maximum of the Perseid meteor shower shifts 2.8 days later each century, so the peak activity now occurs around August 13, but the name given by the Irish hundreds of years ago remains.

Occasionally, very occasionally, a truly exceptional celestial phenomenon is seen, one that literally shouts from the heavens. A certain kind of meteor is like that—one that astronomers call a "fireball." Amazingly, though this only happens very rarely, a fireball may be brighter than the full Moon. Fireballs often cross the sky rather slowly, remaining visible for as long as 5 to 10 seconds. In their celestial passage they often leave a luminous, smoky trail of glowing dust in their wake, and, in exceptional circumstances, it may remain visible for some minutes. Such a phenomenon would have terrified our ancestors and would have been the subject of great astrological speculation for the ancients.

A really brilliant fireball is a rare, once-in-a-lifetime event for any particular point on Earth because these balls are visible only over a very limited area. One was seen in May 1995 in the Canary Islands, the Spanish archipelago off the northwestern coast of Africa, which is now my home. Another was seen in parts of Canada during the summer of 1995. A particularly large and spectacular one crossed the United States in 1972: not only was it seen, but it was also filmed in broad daylight. The most spectacular of all twentieth-century fireballs, however, was observed in a remote region of Siberia on June 30, 1908. This fireball was seen by reindeer herdsmen of the Evenki tribe, in the Tunguska region, north of Lake Baikal. When it exploded, around 4 to 5 miles above the taiga, it was equivalent, in effect, to a hydrogen bomb of as much as 30 megatons. The explosion flattened 40,000 trees over a 42-mile radius but, despite its unbelievable force, it caused no human deaths, thanks to the very sparse population of that inhospitable region.

How large are these fireballs? Let's start with the basics. Most people are surprised to learn that a "normal" meteor is only the size of a grain of dust. A meteor that is the size of a grain of salt would be reasonably bright, and one the size of a grain of sand would appear really bright. On the other hand, a fireball is often as large as a golf ball and perhaps as large as a tennis ball; on very rare occasions it might even be much bigger than this. The inhabitants of the state of Arizona know this because of the famous meteor crater that is such a popular attraction to visitors to the state. I had an occasion to see it one day in May 1996, after I flew out of Los Angeles returning to the Canary Islands. I looked out of the window of the aircraft at one point, and to my considerable astonishment, saw the crater right below us, in the middle of a vast area of completely flat wasteland, with no hills or any vertical relief at all. The contrast between the flatness of the terrain and this huge hole in the ground is quite stunning. The crater is about three-fourths of a mile across and 560 feet deep. There are many differing estimates of when it was formed, ranging from twenty thousand to fifty thousand years ago. Robert Dodd, profesor of mineralogy at the State University of New York at Stony Brook, estimates that the object that caused the crater was an iron meteorite of about a hundred feet in diameter and a mass of 100,000 tons. It would have impacted with a force of around 25 megatons, similar in force to the Tunguska event—one thousand times larger than the bomb dropped on Hiroshima. There are some Indian legends that may apply to this terrible impact, one that would have split the sky with a blinding light before the blast razed everything over a radius of many tens of miles.

Even a piece of material the size of a tennis ball would make a fireball as bright as the Moon, and as such it would take several seconds to cross the sky in a blinding glare of light before probably exploding and crashing to Earth. Had the Magi observed such a fireball? As skywatchers, it would have made a tremendous impression on them—they would never have seen anything like it before. As astrologers, they would certainly have wondered what special meaning it had. What would they have seen? How might it have appeared to them?

Let's imagine the following scenario: the Magi were watching from a tower, late at night. Everything around them would have been still and quiet. They would have been alone with the stars and with their thoughts. Suddenly, a small star would have appeared in the east, where none was present previously. To their surprise, the star would have grown and grown, getting brighter and moving ever-faster across the sky, toward the west. Probably it would have gathered strength for 5 to 10 seconds, becoming impossibly bright to their dark-adapted eyes. The star may even have changed color, becoming bright yellow or orange. As it brightened, pieces would have broken off it like little splinters, trailing in the wake of the star and then fading away as they were left behind. There could then have been a large, silent explosion, after which the star would have disintegrated, briefly getting brighter still. Afterimages would have danced in the eyes of the Wise Men and, as their eyes recovered, they would have noticed a faint smoky-colored trail in the sky where the fireball had been. This trail, the wake of the fireball, would gradually have faded over a few seconds or, in extreme circumstances, over a few minutes.

The Magi would have looked at the sky, amazed at what they had seen. Maybe they were unable to believe their eyes. Perhaps they would have consulted with the elders to seek their opinions, describing to them the exceptional phenomenon they had just witnessed in the sky. Almost certainly none of the elders would ever have seen anything like this star. Horoscopes would be cast to aid in the interpretation of this singular event. The star had come from the east but moved toward the west. Traditionally, a brilliant star would indicate the birth of a king, and the direction of the star's movement would tell them to travel to the west to find him. Jerusalem would have been the obvious and logical place to go.

The arrival of meteors is not an entirely random event. As we have seen with the Perseid meteor shower, they have a degree of regularity. Some eight to ten significant meteor showers can be seen at different times of the year in addition to a host of minor showers. The most important of the major showers are the following ones:

The Quadrantids, seen in early January, which radiate from a corner of the constellation of Boötes, formerly included in the now defunct constellation of Quadrans Muralis (the Mural Quadrant).

The Lyrids, in late April, which radiate from the constellation of Lyra, close to the bright star Vega.
The Eta Aquarids, in early May, which come from the constellation of Aquarius.
The Delta Aquarids, in late July, which also appear to come from Aquarius.
The Perseids, in mid-August, which are the best known and normally the most spectacular meteor shower of the year.
The Orionids, in mid-late October, which come from the north of Orion.
The Leonids, in mid-November, which radiate from the head of Leo.
The Geminids, in mid-December, which occur in the constellation of Gemini, close to the star Castor.

The richest of these annual meteor showers, which appear on the same date each year, can give off as many as one or two meteors every minute under perfect conditions. Other meteor showers may only give one or two meteors per *hour* under perfect conditions and are often almost undetectable even to expert observers.

Meteors have been known for more than two millennia. Chinese and Babylonian skywatchers recorded many observations of meteor showers, some of them identifiable with the showers we see today. Since the tenth century, the Arabs have also kept records of meteor showers. These records report observations that were made mainly in present-day Iraq, where latter-day Arab astronomers followed in the footsteps of the Babylonians. Other observations were made in Andalusia in southern Spain, which was occupied by the Arabs until the end of the fifteenth century.

A meteor shower occurs when the Earth crosses the orbit of a comet. Any comet, in the course of its circuit around the Sun, leaves a quantity of debris behind which, quite literally, makes up its tail or what was the tail of the comet in its previous passes close to the Sun. The stream of dust and debris, cast off by the nucleus of the comet, consists of meteoroids, which become meteors, or shooting stars, when they enter the Earth's atmosphere. These gradually move away from the comet and are then lost forever. But because of the Sun's strong force of gravity, the meteors' orbit is hardly changed when the tail is ejected and hence dust continues around the Sun in virtually the same orbit as the parent comet from which it has separated.

Over the course of thousands of years, the material is spread around the orbit of the comet in a broad band. This band is produced by the force of gravity of the planets and, in particular, by the enormous influence of the planet Jupiter, which perturbs the material and deviates it slightly from its original path. After thousands of years have passed, what was originally a tight stream of material becomes a broad and more diffuse band.

Not every comet produces a meteor shower, because not all comets have an orbit that approaches close enough to the Earth, hence we do not pass through the stream of material that such comets leave behind. The minimum distance within which a comet must pass for the Earth to brush even the outer edges of the stream is around 10 million miles; reduce that distance and the Earth would most probably brush the band of cast-off comet tails or may even plunge deep inside it, causing an important meteor shower. Beyond 10 million miles, our planet would normally not encounter any material at all.

We may still get to see a meteor shower from Comet Hyakutake, the comet that passed quite close to Earth and was such a beautiful celestial object at the end of March 1996. It passed Earth at around 10 million miles and our planet did slip briefly through its tail, though still far away from the head, on March 28, 1996. No meteors have yet been seen coming from the comet, and, with each year that passes, the chance of still seeing them decreases. Comet Hale-Bopp, which was an even brighter object than Hyakutake during March and April 1997, will never show us a meteor shower: various studies have shown that the dense cloud of debris from Hale-Bopp misses the Earth by too great a distance.

Some meteor showers come from quite well known periodic comets. Two meteor showers, the Eta Aquarids and the Orionids, are caused by Comet Halley. By analyzing the distribution of this comet's debris, as estimated by the quantity of meteors seen at different dates and in different years, we can calculate, with a high degree of certainty, that Halley's Comet has gone around the Sun perhaps 3,500 times in its history. This means that it is at least a quarter of a million years old.[1] Thus, even though Comet Halley has been observed for only three millennia, it has been going around the Sun, time and time again, unrecorded by our ancestors, for no less than three hundred millennia. This means Comet Halley has already released an enormous amount of dust out in space, and the comet is now probably only about half of its original size.

Many meteor showers have been known for at least a thousand years, and some have been recorded for as many as two thousand. But it was not until 1834 that any meteor showers were recognized as regular and annual events.

Even so, the Lyrids have been described in chronicles since about 500 B.C.; the Perseids were first recorded by the Chinese on July 17, A.D. 36; the Orionids were possibly observed in A.D. 288; and a big shower from the Leonids was first seen by the Chinese and the Arabs in A.D. 902. Other showers such as the Quadrantids and Geminids are strictly temporary because their orbit is perturbed so much by Jupiter that they continue to cross the Earth's orbit for only a few hundred years. Even though "old" showers die, they are replaced with new ones in the course of time. Arab and Chinese records show a number of meteor showers that are totally unknown and unidentifiable today.

During the times of year when no major showers are active, there are still many minor showers and a continuous drizzle of what are called "sporadic" meteors, which belong to no known shower. These are probably caused by comets that may have disappeared tens or hundreds of thousands of years ago and whose streams of material have dispersed gradually until they are no longer detectable as a meteor shower. Any day of the year, depending on the time of night and the time of year, around ten to twelve sporadic meteors will appear each hour to an alert observer viewing the sky under perfect conditions.

The theory that the Star of Bethlehem might possibly have been one or more meteors has been championed over the years by Patrick Moore, the great British amateur astronomer, writer, and broadcaster. He has proposed the idea in various forms over the years, although the general lines of the theory are very simple. This hypothesis has some attractive aspects, attempting to explain in a plausible way, with a common and natural astronomical phenomenon, why only the Magi seem to have been aware of

Figure 5.1. A meteor seen disappearing behind the VTT solar tower on the right at Teide Observatory. (Photograph by Mark Kidger.)

the new star in the sky. The main suggestion is that the Star of Bethlehem could have been one or possibly two bright meteors (see fig. 5.1), which were seen by the Magi as a "shooting star" during their nights of vigil. On some unknown occasion, while contemplating the sky and pondering its mysteries, the Magi could have seen a bright meteor and reflected on its meaning. The theory is that this first meteor would have sent the Wise Men on their way toward Jerusalem. Some time later, they would observe a *second* meteor while on their journey, which would have directed them to Bethlehem and to the inn when they arrived in the town.

Besides this basic version, a number of different variants on this theme have been suggested over the years.

Most recently, a particularly interesting variant has sprung forth involving the observation of a special and very unusual type of meteor shower called the Cyrilid Stream. A third possible variant on this idea which, to my knowledge, has not yet been suggested in any publication, could be a plausible candidate too. These two variants will be examined later in this chapter.

Ordinary meteors, as we have seen above, are not at all unusual: the Chinese knew about them; the Arabs knew about them. Many records of ancient meteor showers come from Babylon, so we know that Babylonian astronomers were familiar with them. In fact, almost anyone who has watched the sky regularly must be familiar with meteors and with meteor showers. The Magi, we have assumed, were truly wise men who studied the sky. If they were familiar with the planet Venus, as we asserted earlier, then it is equally certain they would have recognized a meteor for what it is—a rather normal phenomenon. On the other hand, a meteor or meteor shower could explain why the references to the Star of Bethlehem are so scarce. As Patrick Moore has commented, "If it had been a bright comet, or a nova, everyone would have seen it; a meteor or two meteors is the only explanation why they didn't."

Such simple reasoning cannot be challenged, but there is another crucial factor to consider: a meteor rarely lasts for more than a second. If we assume, as is more than likely, that the Magi took several weeks to reach Jerusalem, and even further time to plan and arrange their journey, it is implausible that a single meteor could cause them to make their journey, guide them, and finally lead them to Bethlehem. It was this objection that led David

Hughes to conclude in 1976, in an article in the prestigious journal *Nature*, that "the star must have been a fairly long lasting phenomenon, to 'go before' and 'stand over,' which rules out transient phenomena such as fireballs and very bright shooting stars." Although it is difficult for us to understand the Magi's exact motivation for their journey, anything that would make them undertake a long and probably dangerous trip must have been very singular and, probably, very spectacular. But a *normal* meteor makes a very unlikely candidate for a phenomenon to spur the Magi on their voyage, because it would have been a familiar sight to them to see meteors in the sky.

However, let's not be so hasty in rejecting this theory out of hand. What if the Magi had seen something that was not an ordinary meteor? In his comment, quoted above, David Hughes briefly refers to the possibility that the Star of Bethlehem might have been a fireball, but then he rejects the suggestion almost in the same breath. Reconsider the words opening this chapter: a meteor as large as a tennis ball; a fireball as bright as the Full Moon. What might the Magi have seen on that tower, felt as they descended, and reasoned as they discussed the phenomenal sight with their elders? Would they quickly assume a new king was born? The whole theory appears to hold together rather well at this stage, when we look at it this way. Problems start to arise only when we pause to think of how the episode developed afterwards. The Bible appears to suggest clearly that the Star was visible for some considerable period of time—"the Star went before them"—and then says that it stopped over Bethlehem. As brilliant as a fireball might be, and as long-lasting as 5 to 10 seconds might be compared to a normal meteor, the

Bible is suggesting an event that lasted *much* longer than a few seconds. It is simply not possible for a meteor, however bright, to be visible long enough to guide the Magi even at the start of their journey. However well prepared they were beforehand, however much of a hurry they were in, even if they were ready to set out immediately the Star could not have lasted through the journey. The only way that this story of the Star could fit the facts would be if the Magi were already journeying across the desert when the meteor appeared. In other words, they saw the fireball while on their travels and decided, then and there, to follow its path to wherever it might lead.

Moreover, we must consider the inescapable fact that even an extremely brilliant meteor lasting as long as ten seconds would have been a poor guide for the Magi on a journey of several weeks' duration, even if they were already on their way when they saw it. The theory gets around this difficulty, however, by suggesting that a second meteor appeared when the Magi arrived in Bethlehem; this second meteor would have told them when to stop. But here is where the theory starts to lose favor. We assume that the Magi would have traveled mainly before dawn and in the early morning, avoiding the burning heat of the midday Sun in the desert. They would have stopped sometime before midday and would probably have slept through the hottest part of the afternoon. During their weeks of travel, they would have seen many meteors, particularly because one can see more of them shortly before dawn than early in the evening. What would let them know which of the many shooting stars was *the* meteor, the one that was telling them that their search for the newborn king had ended?

We have to assume, to make this theory work, that the second meteor would be another brilliant fireball, just like the first, and that it would appear at just the right moment. If such a fireball is a once-in-a-lifetime event, it is extraordinarily unlikely that *two* brilliant fireballs could be observed, by the same people, at roughly the same geographic point, in the space of just a few weeks. That the second one would appear exactly at the right time as they pulled into Bethlehem is a bit far-fetched. Some observers of meteors, who have dedicated their entire lives to observing the sky, have never seen even *one* really bright fireball.

The Star of Bethlehem had to be something very unusual, but it is unreasonable to stretch the probability of a meteor this far, even though this explanation is theoretically possible. While we cannot *prove* that a meteor is the wrong explanation for the Star of Bethlehem, we can prove that there are other known phenomena at the time of the Nativity that are far better candidates.

Another variant on the meteor theory was proposed in a presentation at the London Planetarium on Christmas 1991.[2] This is a particularly fascinating one and is no less mysterious to astronomers. This theory depends on accounts of a most unusual event witnessed in North America some years ago, the likes of which had never been seen before, nor has it been since.

On this occasion, on February 9, 1913, a most unusual meteor shower was observed. It began with a perfectly ordinary meteor, which spectacularly appeared and slowly crossed the sky trailing a long tail. But this was followed by a second meteor, which curiously appeared in the same position and followed the first across the sky until it disappeared at the same point where the first had previously disappeared. It, in turn, was followed by a third meteor, a fourth, a fifth and more meteors, all crossing the sky in a more and more spectacular procession, one after another, each following the same track in the sky. The whole display lasted for perhaps three minutes.

Were it not for John O'Keefe, of NASA's Goddard Spaceflight Center, the Cyrilid shower would probably have been forgotten. O'Keefe, however, has dedicated time and effort to finding contemporary reports of the shower and analyzing them in detail.

This sequence of meteors was seen from Toronto, in Canada; from New York, Minnesota, Wisconsin, Bermuda, the South Atlantic off the coast of Brazil, and from various other locations on the North American continent. As it was seen on St. Cyril's Day, it was christened the "Cyrilid meteor shower." Professor C. A. Chant of Toronto University, who observed and studied the shower, wrote:

> At about 9:05 in the evening . . . there suddenly appeared in the north-western sky a fiery red body . . . it moved forward on a perfectly horizontal path with a peculiar, majestic, dignified deliberation. . . .
>
> Before the astonishment aroused by this first meteor had subsided, other bodies were seen coming from the north-west, emerging from precisely the same point as the first one. Onward they moved at the same deliberate pace, in twos or threes or fours, with tails streaming behind. . . .
>
> They all traversed the same path and were headed for the same point in the south-eastern sky.

Nothing like it has ever been seen since, and nothing like it had ever been recorded before. It is assumed that it was caused by an ordinary stream of meteors becoming trapped temporarily in orbit around the Earth. As the particles of material swept around our planet, they would have passed through the upper atmosphere, virtually parallel to the Earth's surface.

Around 75 miles above the heads of the astounded observers, each grain of dust in this stream would, in turn, brighten to incandescence. The meteor that we see is not, however, one tiny, glowing grain; it is the trail through the atmosphere that the grain leaves. As it streaks, at anywhere up to 45 miles per second, through the very tenuous gas of the ionosphere that is found at such high altitudes, the heat caused by the friction of the grain's passage rips apart atoms and molecules in the thin air. The tortured atoms lose some of their electrons, leaving a glowing, ionized wake behind the grain: that is the meteor that we see.

The Cyrilid shower has, as far as we know, never reappeared and remains a unique one-time event. Although reported by reliable witnesses, the accounts are not particularly detailed, as is to be expected when ordinary people are taken by surprise by an extraordinary event. For this reason, many of the details of its appearance and behavior are uncertain. This is one of the reasons why we do not have a full explanation for the Cyrilid meteor shower.

Suppose that the Magi observed a meteor shower similar to the Cyrilids. What would they have made of such an orderly celestial display? The insistent motion from one side of the sky to the other would have convinced them to make a journey—the implication would have been irresistible: Follow me! Follow me! The direction of movement would have shown them the way to go. The Magi would never have witnessed anything remotely like this celestial phenomenon before, nor would chronicles, archives, and consultants have had previous records or knowledge of such a thing. In many ways, such an event would be the perfect candidate to be the Star.

This theory, as well, appears quite reasonable and coherent based on the little data we have. But again we run into the same insurmountable fact: it, too, would have been a very short-lived phenomenon. Once again, the Magi would have seen it but would have had no time to react before the stars completely disappeared. We must assume again that the phenomenon, or some similar event, repeated itself several weeks later, like an arrow showing the Magi where to stop and where to find the baby Jesus, unless some other, perhaps completely different event caused them to stop in Bethlehem. Once again, we cannot explain the longevity of the Star that guided them throughout their journey, and once again, we cannot prove that this phenomenon ever existed—nor can we prove it did not.

As with the other meteor theories, we arrive at the same conclusion: this theory is possible, too, but is not one of the more likely ones. Its most serious drawback is that we can never prove it happened because, as it was not recorded, most likely the Magi were the only ones in the whole world to have seen it. Its second fault is the one referred to by David Hughes, that a meteor, fireball, or even a Cyrilid stream is just too short-lived an event to guide the Magi over a journey of several weeks unless it repeated itself later, which is unlikely.

There is one final variant of the meteor theory. It involves an event spectacular enough to be tremendously striking, but of short enough duration that not many people ever see it. It occurs occasionally in the normal course of events: frequently enough to be plausible, but rare enough to have few witnesses. It is a meteor shower that can be checked reasonably easily (if we are lucky) against historical records. This type of rare, highly spectacular event is the so-called meteor storm.

Meteor storms are brief, incredibly intense meteor showers. Instead of one meteor per minute we might observe one meteor per second (a minor storm), or even ten per second (a major storm). The sky literally fills with meteors. They flash by, faster and faster, all radiating from the same point among the stars, appearing as a fiery rain in the sky. On occasion, meteor storms have caused consternation, panic, and terror among those who have seen them (see fig. 5.2), thinking that "the sky is on fire." A typical account is the following from a landowner in the southern United States after the 1866 Leonid shower: "Shortly after 3 in the morning I was awakened by the most pitiful screaming. All the slaves were prostrated on the ground in terror, crying out that the sky is in flames."

A meteor storm occurs when the Earth passes very close to the orbit of a comet, usually close to the time when the comet has passed or is about to pass by. On such occasions, our planet encounters a veritable blizzard of material in its path. The meteors seem to emanate from one point in the sky because the newly liberated material has not had time to spread out and disperse. These fiery lights thus gush forth from a central point in the heavens, filling the entire sky.

Figure 5.2. A Leonid storm. Woodcut. (From A. Guillemin, *Le Ciel: Notions D'Astronomie à L'Usage des Gens du Monde et de La Jeunesse.* Paris: Librairie De L. Hachette, 1870.)

Over the last few centuries, meteor storms have occurred, on average, a few times each century. They are relatively common at present because one particular meteor shower, the Leonids, has caused storms with a certain regularity over the last millennium. Other meteor showers have given rise to occasional storms, but only the Leonids have returned repeatedly with a considerable degree of regularity, as shown in table 5.1.

TABLE 5.1 The Ten Most Spectacular Recorded Meteor Storms

Year	*Shower*	*Meteors per Hour**
1866	Leonids	17,000
1966	Leonids	15,000
1946	Draconids	12,000
1933	Draconids	10,000
1872	Andromedids	7,400
1901	Leonids	7,000
1885	Andromedids	6,400
1799	Leonids	>5,000
1833	Leonids	>5,000
1798	Andromedids	"Like rain"

NOTES: Based on research of the Dutch astronomer, Peter Jenniskens. Five of the ten big storms are from the Leonids. The Andromedid meteor shower no longer exists: it no longer encounters Earth because of the perturbations to its orbit caused by Jupiter. It is possible that the shower may reappear within a few centuries, when further perturbations of its orbit allow the Earth to cross it once again.

*Technically, "Zenithal Hourly Rate."

The Leonid storms occur about every thirty-three years—the time they take to orbit the Sun. For example, a Leonid storm was seen in 1799, others in 1833 and 1866, a lesser one appears, in some North American reports, to have been seen briefly in 1899; but then, one did not occur again until 1966. Other Leonid storms have been registered at roughly thirty-three-year intervals back to A.D. 902. A storm is not necessarily seen every time the thirty-three-year cycle comes up because the band of material that leads to a storm is very narrow: a slight perturbation of the orbit of the meteoric material caused by Jupiter's gravitational pull is sufficient to move the stream of particles away from the Earth enough for the storm not to happen. Though the great 1966 meteor storm was thrilling to witness, it lasted for a mere 40 minutes, illustrating just how narrow this band of material really is. After 1999, Jupiter will perturb the orbit of the Leonid meteor shower severely enough that no new Leonid storm will be seen for at least a century. We will not meet the narrow band of material again until the year 2098, at earliest, and perhaps not until 2131.

Leonid storms follow close behind Comet Tempel-Tuttle, the same comet that causes the weak annual Leonid meteor shower, seen for a few nights around November 17 of every year. Comet Tempel-Tuttle is a faint object with a period of 32.9 years, but it has been seen only five times. One of these times was in 1366, when the comet just happened to pass very close to Earth at a distance of just six million miles. By chance, this encounter, which is one of the closest ever by a comet, was observed by the Chinese over a few nights. A second appearance of the comet was registered in a single observation made with the naked eye by the German astronomer Gottfried Kirch in 1699, though it was not recognized as Comet Tempel-Tuttle until 1965. The third time it was seen, in 1866, it was rather fainter, but still just about visible with the naked eye. On this occasion, it was picked up by those for whom it was named: a German astronomer Ernst Tempel and an American, Horace Tuttle, an extraordinary character who discovered thirteen comets in a highly checkered career that included a five-year spell as a navy paymaster. During the Civil War, Tuttle saw action in the blockage of Charleston Harbor in South Carolina, being credited with the capture of a British blockade run-

Figure 5.3. Comet Tempel-Tuttle during its return in 1998. Although a rather faint object with almost no tail, it was visible with binoculars. (Javier Licandro, with the IAC-80 Telescope at Teide Observatory, Tenerife.)

ner. Ten years later, though, in 1875, he was found guilty of embezzling nearly six thousand dollars of navy funds and was cashiered, although he later returned to work for the United States Naval Observatory in Washington. After 1866, Comet Tempel-Tuttle was lost completely for almost a century before being recovered, telescopically, as a very faint object in 1965.

Comet Tempel-Tuttle returned early in 1998 (see fig. 5.3) and there was reasonable expectation that a new Leonid meteor storm might occur briefly on the night of

November 17–18, 1998, as the Earth crossed close behind the orbit of Comet Tempel-Tuttle. No storm was seen that night, although many astronomers observed a spectacular meteor shower the night before. This shower was later found to have been caused by a return of Comet Tempel-Tuttle in the fourteenth century. It had nothing to do with the 1998 return of the comet. On November 18, 1999, the Earth will cross the comet's orbit once again, and Europeans, probably optimistically, hope to see a meteor storm this time around.

On this occasion, the comet's orbit will once again pass very slightly inside the Earth's orbit, although at a greater distance (792,000 miles) than in 1966, or in 1866, when huge meteor storms were seen. All the recent meteor storms and most of the important showers have occurred when the comet passed just inside the orbit of our planet. It is even possible that, as at the end of the last century, instead of seeing one very big storm we will see several smaller ones; or maybe we will just be unlucky and see nothing unusual. The signs over the last few years, as the Leonids build up to their long-awaited climax, were contradictory, with quite strong Leonid activity in 1994 and 1996, and rather less in 1995.

Some people worry about the dangers that such meteor storms may pose, and, in 1994, NASA even postponed the launch of the space shuttle to avoid any possible collision with a meteor from the Perseid shower. In fact, the dangers are really very slight. Although the meteors appear to be falling like rain and seem to be very close to one another, the actual separation between them is more than 60 miles, even in the greatest meteor storms. Most of the meteors that we see in such a storm are little more than the size of a grain of dust and burn up rapidly in the Earth's atmosphere. A meteorite has never been observed to fall from a normal meteor shower, nor has one fallen from the most intense of meteor storms. Thus there is really very little danger to our planet from the Leonids. There is far more danger to satellites and astronauts from the man-made debris abandoned in Earth's orbit than from meteors. At orbital speeds, even the impact of a tiny fleck of paint could spell disaster. Indeed, quite recently a satellite was damaged in a collision with "space junk." A new telescope at Teide Observatory, Tenerife, where I work, is at present dedicated to finding and mapping space junk.

The danger of an impact with a Leonid meteor does exist for satellites, and, with so many of them in orbit, it is predicted that at least one may suffer an important impact if a big Leonid storm occurs in 1999. No damage to satellites was reported from the 1998 shower, however, and the danger to a single object, such as the space shuttle or the Mir space station, is very small indeed.

Like meteor showers, meteor storms have been recorded many times in history and are nothing new. We know that meteor storms occurred and were already observed more than two thousand years ago. For example, a Chinese chronicle reports that, on March 27, 15 B.C., a great Lyrid meteor storm was seen, and that "stars fell like rain." A meteor storm is among the most spectacular events that can ever be seen in the sky. Few people have ever witnessed one for two main reasons: first, the storms tend to be of short duration, less than an hour, as was the case with the 1966 Leonid storm; second, whether you see the storms or not depends greatly on where you are geographically. For example, American observers saw a huge meteor storm from the Leonids in 1966, but European observers saw little activity just a few hours beforehand. Similar Leonid meteor storms in 1799, 1833, and 1866 coincidentally also had visibility limited to the North American continent.

Could a meteor storm have caused the Star of Bethlehem? Such storms would probably have been unfamiliar to the Magi, and without a doubt would have made a strong impression and had enormous astrological implications. If it *was* a major meteor storm, it could not, however, have been caused by the Leonids; the first recorded Leonid meteor storm occurred in A.D. 902, and there is no evidence that Leonid meteor storms were seen before that date. Another storm, the Lyrids, which is known to have been active around the time of the Nativity, is also an unlikely candidate—not because no Lyrid storm was seen at this time, but because one had already been seen in 15 B.C. Strong Lyrid meteor showers appear to happen approximately every fifty-nine or sixty years, so it is most improbable that one would also have occurred some ten years later, around 5 or 6 B.C.

When considering the idea of a meteor storm as the Star of Bethlehem, we run into familiar difficulties: the Bible speaks of one star, not of hundreds; the storm would have lasted for a very short period of time, at most a single night, and even the most intense part of the storm would have been seen for less than an hour. Once again, it would be hard to imagine that the Magi were guided by such a short-lived event. Furthermore, no major meteor storm was observed at about the right date. If a major storm were still to be found in the ancient records, at a date close to that of the Nativity, it would be an interesting and perhaps a compelling piece of evidence. Without such a record, though, we have to abandon the meteor storm hypothesis.

So where do we go from here? Perhaps a clue to the mystery lies not immediately in the scientific data of sky charts and computerized simulations of the heavens, but rather in a work of science fiction written over forty years ago where the following intriguing sentence is found: "It is strange to think that, before its light fades away below the limits of vision, we may have shared the Star of Bethlehem with the beings of perhaps a million worlds."[3] We will encounter these words again, and more, as we explore new possibilities in the next chapter.

6

Supernova Bethlehem?

WAS THE STAR in fact just that—a star? This seems so obvious, though so far we have looked only at planets alone or in conjunction with others; at comets in their lengthy orbits around the Sun; and at meteors, either singly as fireballs or in showers. In 1956, Sir Arthur C. Clarke wrote a short story called "The Star." In it he assumes the persona of a Jesuit astronomer working as a scientist aboard an exploratory spaceship of the future. The ship's mission is to enter and explore the Phoenix Nebula, the remnant of an old massive star that has died, torn apart by a supernova explosion. In his cabin he tries to reconcile his conscience with his faith, wrestling uncomfortably with the implications of his discoveries.

> We could not tell, before we reached the nebula, how long ago the explosion took place. Now, from the astronomical evidence and the record in the rocks of that one surviving planet, I have been able to date it very exactly. I now know in what year the light of this colossal conflagration reached our Earth. I know how brilliantly the supernova whose

> corpse now dwindles behind our speeding ship once shone in terrestrial skies. I know how it must have blazed low in the east before Sunrise, like a beacon in that oriental dawn.
>
> There can be no reasonable doubt: the ancient mystery is solved at last."

"The Star" was written in 1954 for a short-story competition run by the British newspaper *The Observer*. Much to Clarke's amusement and perplexity, the story was not worthy of even one of the many consolation prizes in the forms of certificates of merit. But when it was later published in the November 1955 edition of the magazine *Infinity Science Fiction*, the story was voted the best science fiction story published that year.[1]

For those unfamiliar with the story, I can thoroughly recommend it. It revolves around a Jesuit astronomer, the central character in the story, who finds that the nebula that he is investigating is the remains of the explosion that was, thousands of years earlier, observed as the Star of Bethlehem. This conclusion forces him into a crisis of faith because of the contradictions involved in the discovery. The contradiction is revealed in the unexpected sting at the end of this classic work, which I will not reveal for the sake of those interested in reading the story.

"The Star" was based on another popular article, also by Arthur C. Clarke, called "The Star of the Magi" and published in *Holiday* magazine in 1954. In this article he suggests that a supernova might have occurred at a distance of 3,000 light years from the Earth. On its death, the hypothetical star's explosion would have been brighter than Venus in the sky and would have burned bright in the oriental dawn. Clarke points out that this explosion would have given it some rather interesting properties:

> The light of Supernova Bethlehem is still flooding out through space; it has left Earth far behind in the twenty centuries that have elapsed since men saw it for the first and last time. Now that light is spread over a sphere ten thousand light years across and must be correspondingly fainter. It is simple to calculate how bright the supernova must be to any beings who may be seeing it now as a new star in their skies. To them it will still be far more brilliant than any other star in the entire heavens, for its brightness will have fallen only by 50 per cent in the extra two thousand years of travel.
>
> At this very moment, therefore, the Star of Bethlehem may still be shining in the skies of countless worlds, circling far Suns. Any watchers on those worlds will see its sudden appearance and slow fading, just as the Magi did two thousand years ago when the expanding shell of light swept past the Earth. And for thousands of years to come, as its radiance ebbs out towards the frontiers of the Universe, Supernova Bethlehem will still have the power to startle all who see it, wherever—and whatever—they may be.
>
> . . . it is strange to think that, before its light fades away below the limits of vision, we may have shared the Star of Bethlehem with the beings of perhaps a million worlds—and that to many of them, nearer to the source of the explosion, it must have

been a far more wonderful sight than it ever was to any eyes on earth.

One of the most popular theories for many years concerning the Star of Bethlehem was that it might have been a supernova. Arthur C. Clarke's suggestion was by no means the first. There was good reason to consider this possibility. A supernova, as far as we know, is the second greatest producer of energy in the known Universe. It is superseded only by the unknown power source that generates the energy of quasars, which are mysterious galaxies located at the edge of the Universe. Quasars look like stars but emit as much energy as tens or hundreds of normal galaxies. Astronomers assume that each quasar is powered by a gigantic black hole, hundreds of millions of times more massive than our Sun. It is thought that a quasar shines on, little changed, for millions of years. A supernova, on the other hand, marks the sudden and violent end to the life of a star—its final and total destruction. Yet in this end, the star offers one of the most impressive and long-lasting spectacles in the heavens. In extreme cases the brightness can rival that of the Moon. More astonishing, though, is that so much light is concentrated into a point of light like a star. For this reason, we take supernovas very seriously as a candidate for the Star, and the Magi would have done the same.

As our technology for investigating the heavens has improved, our theories of what supernovas are have changed as well—in fact, several times in the last few decades; but it seems that the different ideas that have been suggested are finally converging. Now it seems that we really do know what a supernova is, and how it happens. Supernovas are classed into two main types: rather unimaginatively, they are simply called Type I supernovas and Type II supernovas. Some experts suggest that other types of supernova exist, maybe a Type III, or even a IV. They base such theories on small differences from supernova to supernova. For our purposes, we only need to consider the two broader categories. Underlying each supernova type is a different process by which a senile star comes to its violent and final death. The type of star that dies is very different in the two types of supernova.

These days, most of the information that astronomers use to study supernovas and other types of objects comes in the form of spectra. A spectrum is the rainbow of colors that is produced when a beam of light is split up. Superimposed on the rainbow are a series of bright and dark lines called "spectral lines." Each element produces a characteristic set of lines that are as individual to that element as fingerprints are to people. By studying the lines in the spectrum of an object, astronomers can identify its composition. Further analysis and calculation allow the temperature, velocity, and even the density of the gas emitting the spectrum to be measured.

In general, Type I supernovas may seem more powerful than Type II because they emit more visible light. Type I supernovas are further subdivided into two principal subclasses, even less imaginatively denominated Ia and Ib. This division, which is quite recent, became popular only in the past few years as Type I supernovas were found, on detailed examination, to have completely different types of spectra, which suggests they have different origins. Type II supernovas emit less visible light, but are actually more powerful because they emit much more energy. How that is done, we will see later.

Supernovas of Type I

Type Ia

Type Ia supernovas are caused by stars that have already nominally died once, at least nominally, but which have done so in a tired and apathetic way. Their end is a peculiar, three-stage process with a double death that would tax the deductive powers of Sherlock Holmes. As we examine this extraordinary phenomenon, we will not be surprised that the theories explaining supernovas have changed a number of times over the years. In studying supernovas, astronomers try to decipher the mysteries and complexities of events that happened thousands, if not millions, of years ago. Much of what they have to go on is only spectra; hence, the contradictions and false leads. In some cases it is not clear that our ideas are definitive. Some doubts about important details still remain.

But let's try to make some sense of Type Ia. The process that leads up to a Type Ia supernova may take millions of years to reach its final, inevitable, and unbelievably violent climax. Supernovas of this type are found in binary systems, that is, in systems where two stars gyrate endlessly around their common center of gravity. Binary stars are known to be very common, and it is thought that more than half of all known star systems are binaries or multiple stars. In a binary system, one of the two stars is almost always rather larger than the other, often very much larger. Both of the stars are "burning" hydrogen in their centers, converting it into helium through the same reaction that gives rise to the hydrogen bomb. Within the core of the star, four nuclei of hydrogen combine, through various intermediate stages, to form one nucleus of helium. A helium nucleus has a mass slightly less than the four hydrogen nuclei. This difference in mass is released as energy through Einstein's famous equation $E = mc^2$.

In our Sun, 600 million tons of hydrogen nuclei fuse into helium every second, forming about 596 million tons of helium. The difference, 4.3 million tons of mass, is emitted as energy—an enormous quantity of it.

One would expect that the bigger star of a binary system would live longer on its great supplies of hydrogen. The logical thing is that a large star, with much more hydrogen to burn, will last longer than a small star with only a limited supply. But the truth is exactly the opposite. The two stars are, in fact, more like the profligate millionaire and the careful housewife: the profligate millionaire has much more money, but spends it very rapidly; the careful housewife uses her funds with care. In the end, it is the profligate millionaire who runs out of money first.

In the binary system, the large star finishes the hydrogen in its core first. When approximately 7 percent of the mass of the star has been converted into helium "ash," a crisis occurs. As the fuel in the star's core is exhausted, the core stops generating enough energy to support itself. The core then collapses and, because of the compression it suffers, gets hotter, allowing a new nuclear reaction to take place when the central temperature gets hot enough. Helium atoms can combine to form carbon. Three helium nuclei, also known as alpha particles, combine together to form one carbon nucleus in a process known as the "triple alpha reaction" for the three helium nuclei. However, it takes an enormous amount of pressure to force the three alpha particles together, and this means that the temperature of the core must shoot up and its diameter reduce itself tremendously to trigger the reaction. This causes various effects in the star. At this point, things begin to happen rather rapidly, at least in astronomical terms. The outer layers of the star expand. They are blown outward by the tremendous amount of heat and energy generated below them as the core contracts. As a result, the star will turn into what is called a "red giant"—the stellar equivalent of the cream puff: big, but with very little substance.

How exactly does this happen? The reason why this can happen is that most of the mass of the red giant is in the core of the star, which is very small, very dense, and very hot. Due to this high temperature, the nucleus of the red giant produces enormous amounts of radiation. It is the tremendous pressure of this emitted radiation that literally blows the outer layers of the star away. Gravity wants to pull them back, but the radiation pushing outward is stronger than the gravity. Hence the star swells like a balloon. It only stops swelling when the outer parts are so far away from the core that gravity, once more, balances the outward push of the radiation. At this point the star reaches an unhappy and unstable equilibrium, with the star tending to expand and contract as first the radiation and then the pull of gravity dominate in a constant battle of give and take.

Such stars can attain an enormous size. Some red giants, like Betelgeuse, can be as large as or larger than the entire orbit of Mars. But the outer layers of these red giants are very rarefied and, although glowing red hot, produce far less heat than one would expect. The reason is simply that the gas of these outer layers is so tenuous that it is almost a vacuum. It is hot but produces little heat because it has so little substance to it. If one were inside this outer layer of a red giant star, one would seem to be surrounded by a glowing red mist.

When the red giant is in a binary system, a curious effect may occur. If the companion star is quite close to the red giant, material from the red giant may spill over onto the companion. This process is known as "mass transfer." Mass transfer happens when the red giant tries to expand beyond the range of its gravitational influence, known as the Roche Lobe, into the region controlled by the force of gravity of the smaller star. Material from the giant star seeps across this divide, falling onto the secondary star. As the trickle of material crosses over, the secondary star, initially the much less massive of the two, slowly grows more and more massive (see fig. 6.1). Previously we have said that this is a very rapid series of events, and it is, in astronomical time—it may take from only a few tens of thousands of years up to a few million years—just a blink of the eye compared to the ten-thousand-million-year lifetime of our Sun.

In other cases, such as has happened with Sirius, the Dog Star, the sequence goes another way. Sirius has 2.3 solar masses, while its companion, known for obvious reasons as the Pup, has 0.98 solar masses. The Pup was, originally, the much larger of the two stars. When it exhausted its hydrogen fuel and grew into a red giant, it either passed a large part of its mass to Sirius or blew off

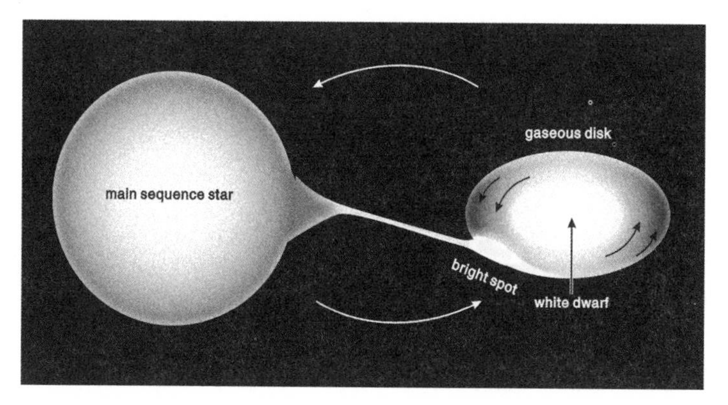

Figure 6.1. A red giant–white dwarf system with the red giant filling its Roche Lobe and transferring mass. (Illustration by Chris Brest.)

most of it when the helium in its nucleus ignited. Either way, so much mass was lost from the Pup that only its nucleus was left behind. This nucleus did not have enough mass to "burn" carbon, so it collapsed to the minimum diameter that the tortured atoms of the broken star, crushed together inside, would permit. The resulting object, of small size and amazing density, fits the mass of a star into a diameter similar to that of a planet and is called a "white dwarf." It is made of "degenerate matter," which is material that is so compressed that it no longer behaves like normal atoms.

When the carbon-rich white dwarf forms after mass transfer, we still have the second star of the binary system, now much larger and brighter, and still happily burning its hydrogen fuel. As its mass has increased, so has its profligate use of hydrogen, sharply reducing its life expectancy. This star will also reach its midlife crisis as its hydrogen core clogs with helium ash, now much more rapidly than when it formed originally. When this happens and it swells into a red giant, the roles of the two stars will suddenly reverse. The formerly smaller star now fills its own Roche Lobe and passes material to the white dwarf. Again, this process might take as long as a few million years.

This time there is an important difference. The stuff that falls is mainly hydrogen from its outer layers. This hydrogen starts to accumulate on the surface of the dead and inert but still very hot white dwarf. This produces an explosive situation because hydrogen can provoke fresh nuclear reactions in the formerly dead star. It will not get to happen in this case only because an even greater catastrophe will overwhelm the white dwarf star first.

Eventually, the weight of hydrogen on the surface of the white dwarf becomes such that a flash reaction occurs. This will happen when the total mass of the white dwarf (in other words, its mass when it collapsed to form a white dwarf star, plus the mass of material from its companion which has fallen onto it) exceeds what is called "the Chandrasekhar Limit." This is the largest mass of white dwarf that can exist without collapsing completely under its own weight. The value of the Chandrasekhar Limit, 1.4 solar masses, is the point at which the pressure exerted by the degenerate matter of the white dwarf (degeneracy pressure), which keeps the star from collapsing further, is suddenly and overwhelmingly overcome by the mass of the star.

The degeneracy pressure is like the foundations of a house designed to be two stories high. If the architect builds the house that the foundations were designed to support, there is no problem. If however, he adds another story, then another and another and another, there will come a point when the foundations cannot support the weight and the building will collapse.

The carbon core of the white dwarf supports an ever-greater mass of surface layers. Like the hydrogen on the surface of the white dwarf, this carbon can suddenly start to burn if it becomes hot enough. As the mass of the white dwarf star passes the Chandrasekhar Limit, the star collapses once again. The core of the star gets tremendously compressed and, at the same time, becomes hotter and hotter. Finally, it causes an intense, very rapid and massive thermonuclear reaction. The carbon core suddenly begins an orgy of uncontrolled nuclear fusion: carbon converting itself into oxygen, neon, silicon, iron, and many other elements. The white dwarf is blown to pieces by an explosion equivalent to the forces of countless billions of hydrogen bombs. This explosion is a Type Ia supernova. Gas from the explosion expands outward in a blast wave that travels at around 6,000 miles per second.

Type Ia supernovas are all very similar to one another, mainly because the critical point at which the carbon-rich white dwarf can take no more pressure is very similar in all supernovas. This means that all supernovas of Type Ia should have almost exactly the same luminosity, and so they can be used to calculate the distance to distant galaxies. If you measure how bright the supernova appears to be and know how luminous it really is, you can calculate the exact distance to it and, thus, to the galaxy in which it appears. Five Type Ia supernovas have been very carefully calibrated using observations from the Hubble Space Telescope, which show that they reach an astonishing visible luminosity that is thirteen billion times that of our

Sun. These are the brightest explosions in the known Universe.[2]

A hypothetical Supernova Bethlehem, if it were a Type Ia supernova, situated at a distance of 3,000 light years, would have shined as brightly as the crescent moon and would have cast sharp shadows at night. If this had happened, it would surely have moved the Magi, both spiritually and literally.

Type Ib

Type Ib supernovas look superficially similar to those of Type Ia and, because of this, they were not recognized as being truly different until the 1980s. A Type Ib supernova, astronomers think, could be a single star, although it is more probable that they, too, are binaries.

The difference between a Type Ia and a Type Ib supernova is that the latter are probably rather massive stars, which pass the helium crisis easily. The star that causes a Type Ib supernova may be between 8 and 25 solar masses. Over the course of its life, it changes from converting hydrogen in its core into converting helium. These stars, however, are sufficiently big that they pass through the first crisis when the helium ash accumulates and suffocates the burning of hydrogen. In a really massive star of this type, there is more than sufficient mass to continue converting helium into carbon.

Even then, the nuclear reactions do not end. As we have seen, in extreme cases even the carbon will react, forming oxygen, neon, magnesium, and silicon. At each stage the core of the star contracts and gets hotter and the temperature reached is great enough to initiate the next stage of nuclear reactions. However, each time this happens, the new supply of fuel lasts for a shorter and shorter time, and the star must take ever more desperate measures to continue the reactions.

When the core of the star is full of silicon "ash," the end is near indeed. In a final attempt to stave off the inevitable, the core of the star collapses once more and reaches a temperature of billions of degrees, sufficient to ensure that it converts silicon into iron. This time there is no escape. Once a core of iron "ash" has been reached, no more stages of burning are possible: no more nuclear reactions exist that would allow the star to make further energy by converting iron into other elements. Any remaining nuclear reactions that may occur with iron, instead of producing more energy, use energy up; iron-based reactions literally suck energy up from the star and make it less—not more—stable.

As the final crisis is reached, the star is built up inside with a series of layers, like the skins of an onion. While the star "burns" silicon in the center, building up a dead iron core, there is a series of layers of other components outside the core. The first of these skins around the core is made up mainly of unburned silicon. In the second skin we find the remaining oxygen, which is being converted into silicon. Then, outside that, there is a carbon shell where the carbon is being converted into oxygen, neon, and magnesium. Then we find a shell, closer to the surface, where helium is still being converted into carbon.

Finally, the last skin of all is a layer of hydrogen, which, at its inner edge, fuses to make more helium. This outer shell of hydrogen comprises more than 99 percent of the diameter of the star, but only about 70 percent of its mass. The hydrogen shell is very extended and, comparatively speaking, is only tenuously held by the force of gravity of the star. If the star has a companion forming a binary system, this companion may suck off almost all of the hydrogen of the outer layers. Even if the star is single, it may lose a large part of the hydrogen to space, blowing off this tenuous atmosphere as what is called a "stellar wind." A stellar wind is just what it seems to be—a stream of gas that blows off the surface of the star, carrying the star-stuff off into space, where it is lost forever.

The star gets to a breaking point as an object that has a core of iron overlain with other elements, but it now lacks the outer covering of hydrogen. When the iron core grows to a certain size, just like the white dwarf star that is made of carbon, it can no longer support its own mass plus that of the different layers above it. The core collapses.

With the foundations of the star suddenly kicked away from underneath them, the layers above the core also collapse. An enormous mass of star, falling inward at very high speed because of the tremendous force of gravity pulling it in, makes an unstoppable force. The whole of the surviving mass of the star crashes into the center in an enormous inward blast wave. In the center of the star, the blast wave from all sides smashes together at the same moment. This blast wave now has nowhere to go but outward, bouncing back from where it came. After crashing together from all sides in the middle of the star, the blast wave rebounds outward, destroying the star with a shock wave of unbelievable violence and leaving behind a glowing cloud of expanding wreckage.

What would you see if a relatively nearby star suddenly

turned into a supernova? Probably one would not notice anything out of the ordinary until the explosion actually happens. Until then the star would look quite normal. From a safe distance—which, for the Earth, even protected by our atmosphere, is a minimum of thirty light years if we wish to avoid being fried by radiation—it would be quite a show. As the explosion gathers force we would see how the star increases in brightness very rapidly. In a few hours it would be at least ten thousand times brighter than before. You could see the star getting brighter as you watch. Even if the explosion were to happen just a few thousand light years away, we would see it in broad daylight. The star would remain visible in daylight for several weeks. Initially, through a telescope, the explosion would just be a point of light. But, over the following months, as the light from the explosion fades, the glowing cloud from the explosion would become visible. Within a few years the star itself would have faded from sight, but the expanding cloud of wreckage around it would be visible with even a small telescope.

Supernovas of Type II

Apparently less spectacular than a Type I supernova are the supernovas of Type II. As is often the case, looks can deceive. Type II supernovas seem less violent but are actually more violent and more energetic than supernovas of Type I, even though they may appear less spectacular because less of their energy is emitted as visible light.

The most massive stars of all are the ones that will eventually convert themselves into supernovas of Type II. The process is very similar to that of Type Ib but more

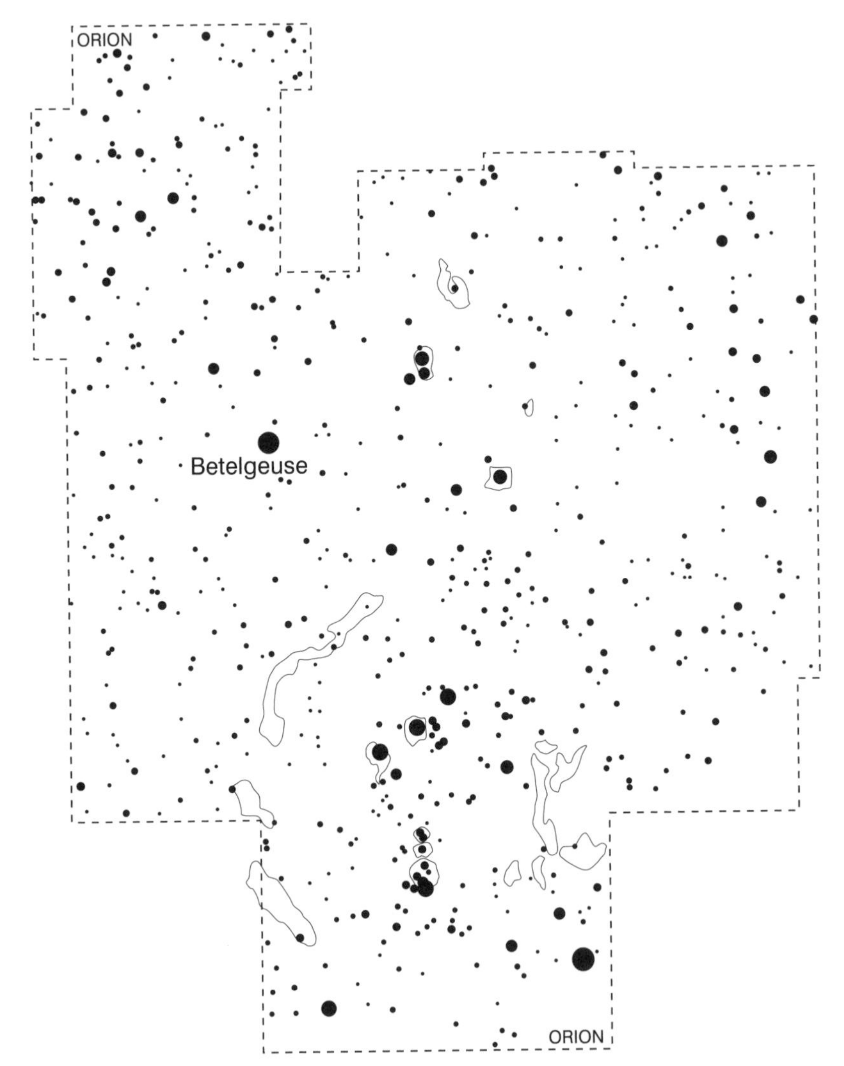

Figure 6.2. Situation of Betelgeuse in Orion.
(Illustration by Chris Brest.)

extreme. A typical star that will one day turn into a Type II supernova is the well-known, first-magnitude star Betelgeuse. Betelgeuse, which marks the left shoulder of Orion, is one of two principal stars of the constellation and one of the brightest stars in the sky (see fig. 6.2). It is called a "red supergiant"—in other words, it is a real giant even among red giants. Betelgeuse is some twenty times the mass of our Sun, but it is around eight hundred times the Sun's diameter. Its diameter, nearly four times that of the Earth's orbit, is so great that, were it placed in our solar system, all the planets from Mercury through Mars and part of the asteroid belt would be within its globe.

Betelgeuse is very close to the end of its life. Within its bloated surface we can find layer after layer of different nuclear reactions, down to the core, where silicon is being fused into iron (fig. 6.3). The star is unstable. Its instability is visible for all to see because the star is pulsating, expanding in and out with an irregular cycle and getting alternately brighter and fainter as it does so. Here we are witnessing the battle between the star's force of gravity, which tries to suck these outer layers in, and the pressure of the radiation emitted within it, which is trying to blow the outer layers away. Curiously, Betelgeuse is actually brightest when it shrinks, because the surface of a star is hotter when it is smaller and, despite its smaller size, it will emit more light than when it has swollen to its maximum diameter and is cooler. As this tug of war between gravity and radiation pressure continues, Betelgeuse occasionally fades to well below magnitude 1; at other times, though, it may reach magnitude 0, making it one of the brightest stars of the entire sky.

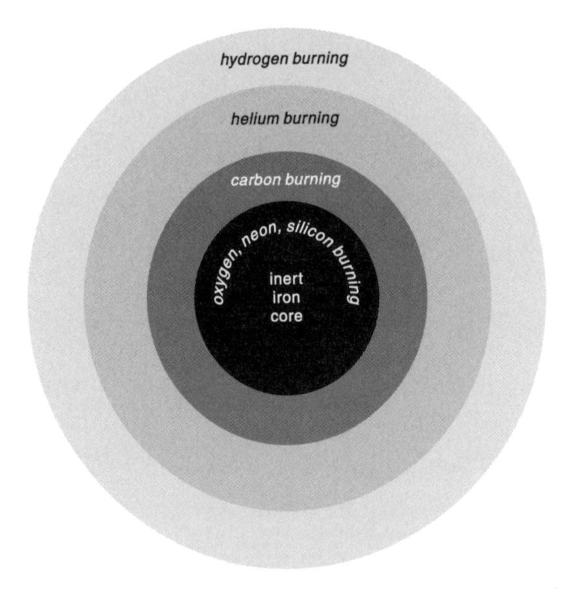

Figure 6.3. The interior of a red supergiant showing the "onion layers." (Illustration by Chris Brest.)

This is possibly a good place to discuss how astronomers measure the brightness of stars. They use a scale known as the magnitude scale. This is based on the system introduced in the second century A.D. by the Greek astronomer Ptolemy, who assigned a magnitude to each star in the sky, calling the brightest stars magnitude 1 and the faintest ones that he could see with the naked eye, magnitude 6. We still use this same scale with some small variations today.

The magnitude scale is defined in such a way that a star that is five magnitudes brighter than another is exactly one hundred times brighter. The magnitude scale goes in reverse, like a golfer's handicap. The very brightest stars have a negative magnitude, while the faintest ones that can be detected by the largest telescopes are around magnitude +28. If a star has a negative magnitude, it is very bright. Similarly, if a golfer has a negative handicap, he is a brilliant player. In contrast, if your handicap is +28 you have a problem with your game; likewise, a star of magnitude +28 is a very faint one indeed.

The difference between the very brightest star in the sky, Sirius, which has magnitude −1.4, and the faintest stars detectable with the Hubble Space Telescope is about thirty magnitudes, or a factor of 1,000,000,000,000 (a trillion) in brightness.

It was John Herschel (1782–1871), son of Sir William Herschel, the discoverer of the planet Uranus, who discovered the variability of this star. John Herschel, an important astronomer in his own right, noticed how, on November 26, 1839, Betelgeuse appeared to be brighter than Rigel in the constellation Orion. Normally, Betelgeuse is clearly the fainter of the two, so this is in itself an unusual event. Rigel is magnitude +0.08, and thus Betelgeuse was at least equal to or more brilliant than magnitude zero at that time.[2]

Less than a month and a half later, on January 7, 1840, Betelgeuse was now fainter than Aldebaran, which is a star of magnitude +0.86 and, hence, must now have been around magnitude +1 or even fainter. Modern observations of this star show that such a large variation in such a short period of time is most unusual, not to mention virtually unknown. John Herschel was thus fortunate to witness something totally out of the ordinary.

However, the strange antics of this star were not yet over, and John Herschel remained vigilant. On December 5, 1852, Betelgeuse became even brighter, clearly outshining both Vega and Capella. Normally Betelgeuse is fainter than both these stars, but brighter than Aldebaran, hence it is usually around magnitude 0.5. In 1852 it must have become significantly brighter than magnitude 0. Over the last few years it has been more stable in brightness, but sudden changes in its intensity may still occur, as happened in the winter of 1994–95. Some have suggested that the star has an approximate five-year period, but this is likely to be the result more of wishful thinking than of Betelgeuse's own behavior.

We do not know how far Betelgeuse is from the crisis point, but we know that it cannot be far. Iron is accumulating in its core and suffocating the nuclear reactions that support the mass of the outer layers. We do not know and will not know how far advanced the process of suffocation is until the crisis finally occurs. That said, Betelgeuse probably has, at best, a few thousand years to survive, and there is even a small chance that it has already exploded. As this star is approximately five hundred light years away from the Earth, we are seeing it now as it was around the end of the fifteenth century. All we can say is that, by the time Columbus arrived in the Americas, nothing had yet happened to Betelgeuse, because, if it had, the news would, by now, have reached the Earth. It is possible, although unlikely, that Betelgeuse already exploded three or four hundred years ago but that the light from the explosion has not yet reached us.

At maximum, supernova Betelgeuse will be about as bright as the Full Moon and would be expected to remain visible to the naked eye for more than ten years. The night will be illuminated for weeks by a blindingly bright point of light. Imagine all the brightness of the Full Moon concentrated into a single point. The intensity (or "surface brightness") of the light would be exactly intermediate between that of the surface of the Moon and of the Sun, if we allow for the fact that their light is spread across a large area of sky.

Even though supernova Betelgeuse will appear to be two magnitudes fainter than an identical supernova of Type I, its true power is not in visible light. Ten times as much energy as the visible light is released as kinetic energy in a Type II supernova. The motion of the huge gas clouds emitted by the supernova at several thousand miles per second carries away fantastic amounts of energy. Even greater than the kinetic energy is still a third energy source. A huge amount of the total energy from the explosion, as much as 99 percent, is released in such a way that it is almost totally undetected and undetectable. Although it seems incredible, this vast quantity of energy is emitted as the tiny phantasmal particles known as neutrinos.

Galactic Supernovas

If the Star of Bethlehem was a supernova, it must have been one that appeared within our own galaxy, as supernovas in other galaxies would be too far away and too faint to be visible with the naked eye. Supernovas within our galaxy are quite rare; older books quite commonly state that only three have been seen in the last thousand years. The number is actually at least four and probably even more (see table 6.1).

The brightest supernova ever observed was the one seen in the constellation of Lupus the Wolf in 1006. This supernova appeared so far into the southern sky that it would have been below the horizon in much of Europe

TABLE 6.1 Probable and Possible Supernovas Observed within Our Galaxy during the Last Two Millennia

Year	*Constellation*	*Magnitude*	*Duration*
185	Centaurus	−8	20 months
386	Sagittarius	3?	3 months
393	Scorpius	0	8 months
1006	Lupus	−9.5	2 years
1054	Taurus	−4	21 months
1181	Cassiopeia	0?	6 months
1572	Cassiopeia	−4	18 months
1604	Ophiuchus	−2.5	12 months
1680?	Cassiopeia	6?	1 night

NOTE: Many more supernovas have exploded unseen or unrecorded.

and China, and it would barely have skirted the horizon from Beijing, or from the south of central Europe. The most valuable observations of this supernova come not from China, but from Saint Gallen in the south of France, where it could only be seen briefly each night through a gap in the mountains on the southern horizon. The supernova of 1006 has left a small, circular radio source similar to a broken ring. Observed through an ordinary telescope, this expanding ring of gas is almost invisible and barely detectable. It was almost certainly a Type I supernova at about 4,000 light years distance, a relatively near neighbor in space. This supernova has been accorded the status it deserves only in the last few years.

The most famous of the supernovas is, without a doubt, the one that occurred in 1054. Often known as the Crab Supernova, it gave rise to the spectacular Crab Nebula (see fig. 6.4), which can be seen with a small telescope and was visible in broad daylight for twenty-three days. The Chinese

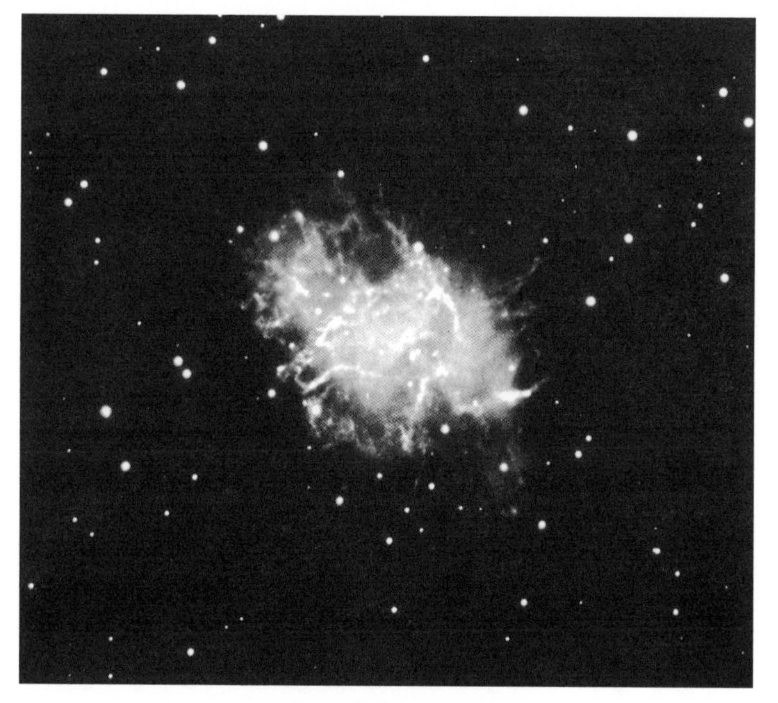

Figure 6.4. The Crab Nebula.

observed it thoroughly and carefully, but, surprisingly, it appears not to have been recorded in Europe apart from a single, indirect mention in one chronicle. This single reference was by Ibn Butlan, a Christian physician from Baghdad who eventually became a monk. However, the Crab Supernova is recorded on many primitive rock paintings found around the United States, showing a crescent Moon and a star beside it: on July 5, 1054, the supernova would indeed have been very close to the Crescent Moon. To an observer in the eastern United States, the waning Crescent Moon, just 10 percent of its disk illuminated, would have passed just 3

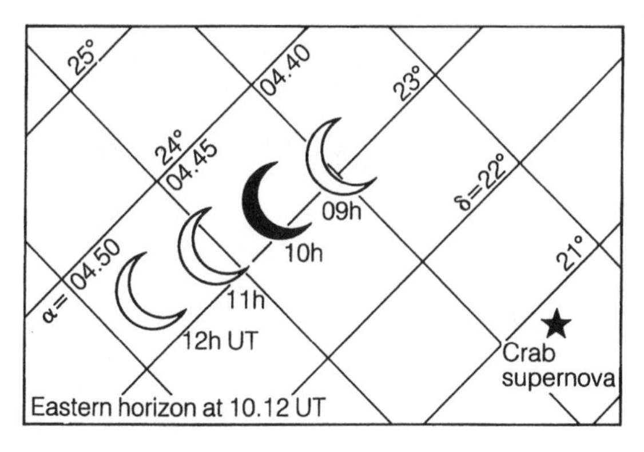

Figure 6.5. The moon and the Crab Supernova. (From Paul Murdin and Leslie Murdin, *Supernovae*. Cambridge University Press, 1985, pg. 14. Reprinted with the permission of Cambridge University Press.)

degrees above the supernova around dawn. The conjunction of the two objects would have been tremendously spectacular (see fig. 6.5). The Crab Supernova, which occurred 6,000 light years away, was very probably of Type II. It was much fainter than the supernova of 1006 even though it was only a little farther away, and it had the slower decline that typifies a Type II.

The supernova of 1572 is the best observed of all Galactic supernovas, thanks to the Danish astronomer Tycho Brahe (1546–1601), who studied it in detail. For this reason it is usually called "Tycho's Star," although Tycho was certainly not the discoverer. Over the supernova's year and a half of visibility, Tycho measured its position, brightness, and color with a precision that was centuries ahead of its time. His observations yielded more detailed information than is available for many recent supernovas and has proved both valuable and highly reliable. Today the remnant of the supernova can be seen optically through a large telescope, but it appears faint and wispy. Examined with a radio telescope, Tycho's supernova is a very powerful source of radio emission, forming an almost perfect ring, as the cloud of gas expelled by the explosion expands outward like a ripple on a pond. Tycho's Star was a Type I supernova situated at about 20,000 light years distance from the Sun.

The supernova of 1604 also had a distinguished observer in the German astronomer Johannes Kepler (1571–1630), a student of Tycho Brahe. Tycho himself had died in 1601 and thus just missed the chance to observe two bright supernovas in his lifetime. Once again, the person whose name is associated with the supernova, Kepler, was not the one who actually discovered it. Justice is better served in this case because Kepler was, quite genuinely, one of the first people to see it, which Tycho could not claim of the 1572 supernova. Like his distinguished tutor in 1572, Kepler made sure that he measured it with all due care and attention to detail, particularly its variations in brightness. This supernova was much less bright than Tycho's because it was more distant and partly hidden by the dust clouds that lie in front of it. Its expanding cloud of stellar wreckage is still a bright radio source, though the star is rather faint in visible light but easily detectable with large telescopes. As with Tycho's Star, it was a Type I supernova, and estimates of its distance range from 20,000 to 30,000 light years. One curious difference with other supernovas of the last millennia is that this one appeared close to but possibly slightly above and behind the center of our Galaxy, while the others (except perhaps Tycho's supernova) have appeared far from the center, in nearby spiral arms.

But let's travel farther back in time for evidence of a possible supernova Bethlehem, though the three most ancient candidates to have become supernovas are more difficult to prove. A star observed in A.D. 185 was mentioned only briefly in a single Chinese chronicle that was written more than two centuries later. Today a powerful radio source lies close to its position. Almost as convincing is that, for the star mentioned in the chronicle to have been seen at all, when it was very low in the dawn sky of December 7, A.D. 185, it must have been exceptionally brilliant. This brilliance and its long duration of visibility suggest it must have been a supernova.

A star seen in A.D. 393 was also bright enough and long-lasting enough that it was most likely a supernova, though it is not possible to confirm its identity with the radio source that appears to be the result of this supernova. The other object, from A.D. 386, is even more doubtful: it could have been a faint and distant supernova, and a nearby radio source looks about the right age and type for such a star, but it is not a conclusive case.

The suggestion that the Star of Bethlehem might have been a supernova is not new. The idea that the death of a star, blazing in the sky above the Holy Land, may have signified a new birth seems in some way right and appropriate. In fact, when Tycho's Star was seen, Renaissance Europeans commonly interpreted it as a recurrence of the Star of Bethlehem. The reason for this belief in the recurrence of the Star is that Tycho's Star surprised and bewildered the masses. They were used to the view that the stars were permanent and unchanging, but this "new" star was quite spectacular and must have held special meaning for them. Tycho did not share this interpretation, even though he did not really know the nature of the star he had observed. He suspected that it was distant, but could not imagine just how far away it really was. Even so, his failed attempts to measure its parallax and movement in the sky helped to convince him that the old model of the Sun, Moon, planets, and stars set in crystal spheres was untenable. He saw a new star and left it at that, never suspecting that it was the funeral pyre of an old and tired star, 20,000 light years away.

Is it possible that it was a supernova that shone over Bethlehem? At the time of Tycho, the only known references in Europe to ancient observations were those in the Bible and in a few Latin and Greek texts. The priceless archives of Chinese chronicles and, to a lesser extent, the Korean and Japanese observations, which we will look at later, were wholly unknown. Even in the mid-1950s, the extent of the information in those largely untranslated archives was still not known. When Arthur C. Clarke proposed the supernova theory, it was entirely possible that a supernova could have appeared around the time of the Nativity, although it would have been unprovable at the time.

What would the signs of a supernova be? To find out, we have to search for an observed bright "guest star." A faint object, like the possible supernova observed in Sagittarius in A.D. 386 would not fit the bill. To be a good candidate, it should have been observable for a long time. A Type I supernova would fade by seven magnitudes, a factor of 630, in around three hundred days; a Type II supernova would fade the same amount in around four hundred days. A bright supernova would thus be visible for at least nine months and probably for more than a year. The other requirement is that it had to be an object that was not recorded as either moving in the sky or as having a tail—both sure signs that the object is a comet and not a supernova. This strong set of criteria rules out most astronomical phenomena, and indeed, it rules out a supernova as the Star of Bethlehem. Our knowledge of the Chinese chronicles has improved to a great extent, as most of the important Chinese observations have been located and translated. None of the objects observed by the Chinese as having appeared around the time of the birth of Christ fit the criteria for a supernova.

The supernova closest to the date of the Nativity was that of A.D. 185, but of course it came nearly two hundred years too late. No object seen between 15 B.C. and A.D. 10 was observed for more than three months. At least one of the few objects that were observed during that time was a comet. Even if the Chinese, for some reason, had failed to register the object, we would expect to find a strong and characteristic radio source in the sky today, but none is seen that appears to be a supernova remnant about two thousand years old.

However, neither the absence of the Chinese record nor the absence of a radio source is totally conclusive on its own. In some cases, as we have seen, we cannot tell for sure which of the several sources in a crowded region of the sky is the supernova remnant that goes with a particular supernova. Though the Chinese chronicles contain some mistakes and omissions, they have usually been corrected and filled in by the Korean chronicles and, to a lesser extent, the Japanese. For this time period, however, there is no evidence for a "missing" supernova, and the reason is clear and evident: there was no supernova to see. Despite the attractiveness of the supernova hypothesis, it fails to meet the most basic and fundamental criteria.

I will return in chapter 9 to the Chinese, Korean, and Japanese chronicles, the extraordinary records that take us way back to ancient skies and astronomical events. So impressive are these chronicles and so important are they to our search for the Star that they are worthy of a whole chapter. But first, let us travel to ancient lands and peoples and get to know them. We have spoken of the Magi throughout the book, but who were they really? Where were they from? And why did they follow a star?

7

We Three Kings?

IN THIS CHAPTER, we return to a topic we first took up in chapter 1. This retreat is necessary. In the last few chapters we have explored some quite astonishing astronomical phenomena, a number of which could have been the Star —if the timing was right, if this, if that. But none seem to fit our criteria for being the Star, so we have arrived here without seeming to have advanced much. Up to now we have seen mainly what the Star could have been but almost certainly was not. Rather than continuing to search the heavens once more with telescopes that can "see" stars that died thousands of years ago, let's look back in time here on Earth and examine clay tablets, cave drawings, and other records. If the Star was seen and if we can understand those who are said to have seen it, then we will better understand what phenomenon might have spurred them into their journey across the desert. Who were these Wise Men, or the Magi, to whom Matthew refers in chapter 2:1–2.

> In the time of King Herod, after Jesus was born in Bethlehem of Judea, wise men from the East came to Jerusalem

> Asking, "Where is the child who has been born king of the Jews? For we have observed his star at its rising, and have come to pay him homage."

Today, Christmas carols and Nativity plays immortalize the Magi in various ways. In the English-speaking world, millions of children have learned the words,

> We three kings of Orient are,
> bearing gifts we travel afar,

which open one of the most famous of all Christmas carols in the English language. In Nativity plays and in popular culture, these three people, of whom we know so little, are called Melchior, Balthasar, and Gaspar. In the popular tradition of many countries, one of the three is dark-skinned.

In many Catholic countries the arrival of the three "kings of Orient" is awaited with much greater anticipation than is Father Christmas, or Santa Claus, or Saint Nicholas.[1] Spanish television, for example, dedicates enormous coverage to the arrival of the three kings. They are seen entering various towns and cities of Spain on the evening of January 5, and it is quite normal for three of the four national channels to carry simultaneous live coverage of the event. In many cases, the kings are greeted as foreign dignitaries by the mayor and other officials and then parade around the town on camels, reassuring the children that they have arrived safely and will soon turn to the work of leaving gifts for them during the night.[2]

The best way to find out who the Magi were and where they might have come from is to answer a number of questions about them, based on our traditional image of them. As we will see, much of the modern tradition of the three kings from the east comes from shaky ground, at best.[3]

Who Were the Magi? Were They Really Kings?

The truth is that we know next to nothing about the Magi. In the traditional translations of the Bible they are usually referred to as Wise Men, although a recent innovation is to refer to them as astrologers, as in the New English Bible. In the New Revised Standard Version, "astrologers" is given as an alternative to "Wise Men" in the footnotes.

The word "Magi," which is commonly used to describe them, is an anglicized version of the original Greek word used in the biblical text, *magoi* (μαγοι), the plural form of *magos* (μαγος). (See the further discussion of the word later in this chapter.) In Spanish, this word has been traditionally translated as "Reyes Magos," literally, "Magician Kings." Although Tertullian, one of the early church fathers who was brought up in Carthage, Africa, referred to the Magi as kings as early as A.D. 250, it was not until the sixth century A.D. that the Wise Men were transformed into royalty in the mainstream church. It was still later, around the tenth century, before the practice of referring to them as kings became widespread. The representation of the Magi as kings is thought to have been almost certainly related to the politics of the early church, where the kings of men were regarded as being such only because they were subjugate to the King of Kings.

It is also possible that this royal interpretation comes

in part from the Old Testament prophecies about the birth of the Messiah, or from passages that can be regarded as referring to his coming. For example, Psalms 69:29,

> Because of your Temple at Jerusalem kings bear gifts to you.

Or Psalms 72:10,

> May the kings of Tarshish and of the isles bring him tribute, may the kings of Sheba and Seba bring gifts.[4]

Or even Isaiah 60:3 and 60:6,

> Nations shall come to your light, and kings to the brightness of your dawn.
>
> . . .
>
> A multitude of camels shall cover you, the young camels of Midian and Ephah; all those from Sheba shall come. They shall bring gold and frankincense and shall proclaim the praise of the LORD.

Note how closely the passage from Isaiah, which would have been well known to Matthew, corresponds to his description of the visit of the Magi. As it was prophesied here that the Messiah would be visited by kings, it is not difficult to imagine that later writers "corrected" Matthew's account to include this important detail and converted Matthew's Wise Men, or Magi, into royalty.

Some commentators note that the biblical account of the rather low-key visit of the Magi does not really square with the arrival from distant lands of three exotic and important kings and their entourages. If the Magi had been inportant personages, their arrival in Jerusalem and in the sleepy little village of Bethlehem would surely have been more widely commented on. Thus the argument that the Magi were *kings* could not be true, and there is no reason to believe they were royal visitors.

Were There Really Three Magi?

The Western tradition that there were three "kings" is not original, nor is it based on firm evidence. Furthermore, *three* kings, rather than another number, were assumed simply because three gifts were offered: gold, frankincense, and myrrh. Augustine of Hippo, around the end of the fourth century, writes as if the three-Magi tradition were accepted by the whole gentile world. But the earliest drawings of the Nativity, found in the Roman catacombs, show four Magi in some drawings and only two in others. In addition, not everyone accepts the assumption that there were only a small number of Magi. In the eastern tradition there were twelve kings. In the *Chronicle of Zuqnin*, an eastern work that was finished in the eighth century, a story recounts how the twelve Magi gaze at the Star and each sees a different image reflected in it.

Nowhere in the Bible or the Protoevangelium is the number of Magi mentioned; we can only affirm there was more than one because the texts clearly use the plural to describe them.

What About Their Names and Origin?

The traditional names of the Wise Men in Nativity plays are Gaspar, Balthasar, and Melchior. These names are of more recent origin, and are not mentioned either in the

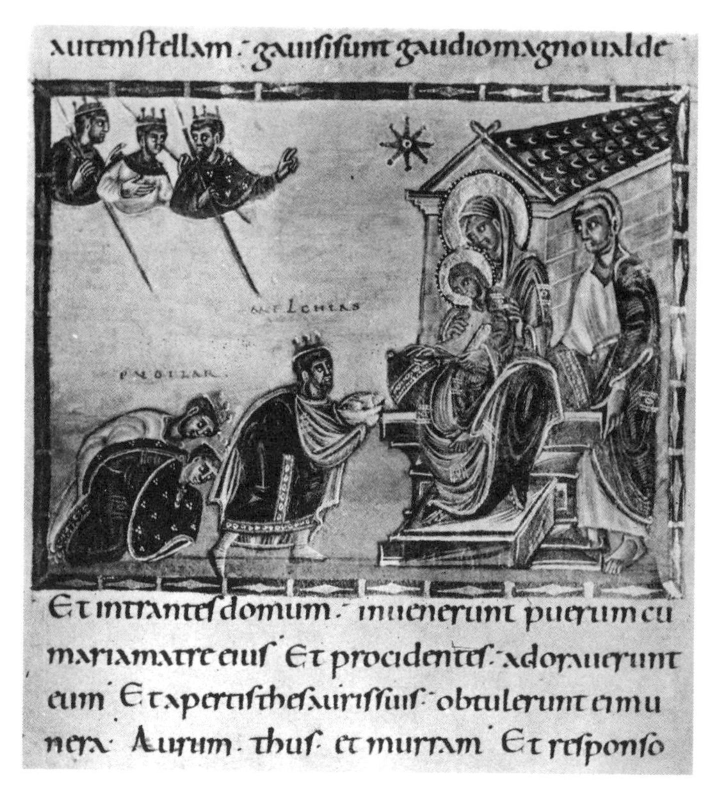

Figure 7.1. *Adoration of the Magi,* Codex Egbertus. (Foto Marburg/Art Resource, NY.)

Bible or in the Apocrypha. The Reverend Phillip Greetham has pointed out that some accounts of the Nativity give completely different names: in one they are Hormizdah, Yazdegerd, and Perozdh, in another they are referred to as Hor, Basanter, and Karsudan.[5] The names Gaspar, Balthasar, and Melchior date back to Origen in the third century A.D. and became popular by the sixth century (see fig. 7.1). Trexler disagrees with this chronology, stating that these names did not appear until the fifth century, long after the death of Origen, and they did not become popular until the end of the first millennium.

Some accounts of the Nativity suggest that at least one of the kings was a Chaldean of dark skin, but this too is a comparatively modern addition (see fig. 7.2). This claim lends itself to the interpretation that the Magi may not all have come from the same place, a possibility that is often overlooked. In fact, in some versions of the Nativity dating back to the end of the first millennium, the three Magi represented three completely different races and points of origin, with Melchior a dark-skinned African, Gaspar a fair-skinned European, and Balthasar an Asian.

Trexler, in his elaborate and highly detailed story of the development of the image of the Magi over the centuries, points out that, in the late seventh or early eighth century, the Venerable Bede, one of the few scholars and historians active in Saxon England during the Dark Ages, wrote about the Magi in the following terms: "Mystically, the three magi signify the three parts of the world, Asia, Africa, Europe." This development of the story seems to have been lost in the more modern versions of the Nativity, which usually assume tacitly that all three Magi came from the same place and traveled together.

In different accounts, however, the races of the three Magi are changed. The most curious anecdote about the Magi is recorded by Trexler. The story goes that in 1610 an Ethiopian named Joan Balthezar visited a Dominican named Luis de Urreta in Valencia, Spain, and claimed to be a descendant of "King" Balthasar.[6] He further claimed that descendants of Gaspar and Melchior were also well and alive in Ethiopia, that Gaspar was originally a native of the country, whereas Balthasar and Melchior had aban-

Figure 7.2. *Adoration of the Magi*, detail, by Jan Swart van Groningen. (Bob Jones University Collection.)

doned their former homes in Persia and Arabia to escape persecution by non-Christians and also settled in Ethiopia. Joan Balthezar ended his story by adding that rule of Ethiopia had passed from one family of descendants of the Magi to another, in strict rotation.[7]

Despite some quite obvious holes in Joan Balthezar's account, it seems that Luis de Urreta was completely convinced by it. Urreta soon suffered the indignity of seeing his gullibility ridiculed in public, particularly by a Jesuit named Pedro Paez who consulted the ruler of Ethiopia about Urreta's version of Ethiopian affairs. The Ethiopian ruler, unsurprisingly, debunked the whole story and denied vigorously both that he was descended from the Magi and that the throne rotated among Magi families.

Thus, even the names of the Magi—Gaspar, Melchior, and Balthasar—are no guide to their identity, their number, or even their existence.

What About the Gifts They Bore?

As we have seen, Isaiah prophesied that gold and frankincense would be given to the newborn king as gifts. These, in combination with myrrh, are highly symbolic items. Gold has long been regarded as one of the symbols of royalty and was an obvious offering. It was mined or panned in eastern Egypt, Western Arabia, Armenia, and Persia, and was widely available to those rich enough to import it. One legend has it that the gold the Magi offered came from a cave where Adam stored some of the wealth of paradise. Over the centuries, different churches would claim that their chalices were made from this selfsame gold, which had been given to Mary and the infant Jesus.

Frankincense is a yellow gum that comes from the frankincense tree, *Boswellia sacra*. It is "milked" from trees in much the same way as rubber or maple syrup. The bark is peeled away and the wood underneath is cut. In time, a yellow liquid oozes out of the wound, solidifies slowly, and gives rise to a resin that gives a sweet scent when heated or burned. Frankincense was traditionally used in holy anointing oil and thus symbolizes the priestly office of Jesus.[8] By giving him frankincense, the Magi were acknowledging that Jesus would be an important prophet or preacher.

Myrrh is prepared in a similar manner to frankincense, though it comes from a shrub, *Commiphora myrrha*. This plant, found in the Horn of Africa and in the southern part of Arabia, exudes a yellow-brown oily resin that is highly fragrant. Although it was also an ingredient of holy anointing oil, its symbolism here is more specific: it was used to anoint the dead. The gift of myrrh would thus symbolize the way that Jesus would later die on the cross.

The three gifts are thus highly symbolic: two of them, as prophesied by Isaiah, represented the ministry of Jesus as the newborn "king," the third symbolized his death, which would turn into one of the most important cornerstones of the Christian faith.

What Was a "Magus," Anyway?

There is good reason to believe that the word "magi" refers to a priestly caste that interpreted observations and prepared horoscopes. This definition would tie in with the translation "astrologers" used in some versions of the Bible. Here we have some degree of agreement among various writers.

David Hughes, astronomer and astrohistorian, is absolutely clear in his conclusions: "They [the Magi] were probably Median priests of Zoroastrianism who, apart from performing the duties of national priesthood, used to occupy themselves with the interpretation of dreams."[9] Hughes also suggests as an alternative that "Magi" was simply the general name of a priestly caste that had "magical" tendencies. Colin Humphreys largely agrees with this analysis.[10] He quotes the historian Herodotus—who was born about 485 B.C. at Halicarnassus, home to one of the Seven Wonders of the Ancient World—as saying that Magi existed in Persia in the sixth century B.C. as a religious group among the Medes, where they carried out religious ceremonies and interpreted signs and portents.

The Reverend Greetham adds that the word "magus" was originally associated with the Medes and Persians and dates back to the Persian prophet, Zoroaster. Precisely when and where Zoroaster, whose name may come from the Greek words meaning camel handler, lived is a matter of some debate. Some suggest that the religion he founded, Zoroastrianism, stems only from the sixth or seventh century B.C.; others claim it is far older, going back at least to 1000 B.C. or even as far back as 1700 B.C. In identifying the magi with the Medes, Greetham is in agreement with Hughes and Humphreys, but he adds one further significant detail: according to Greetham's research, the word "magi" did not come into common and popular usage until the first century A.D., some time after the Nativity.

In the Old Testament we find many mentions of "magicians" (from the same root as "magi"), particularly in the book of Daniel. To cite just one example, Daniel, chapter 2, describes the way that the "magicians, the enchanters, the sorcerers, and the Chaldeans" of Babylon were summoned to King Nebuchadnezzar to interpret his dreams,[11] and how Nebuchadnezzar decided to execute and "destroy the wise men of Babylon" because he was less than happy with their interpretations.[12] It does not take a great leap of the imagination to link these magicians, or wise men, with the Magi, or Wise Men, of the Nativity story.

However, the word "magi" is also used in the New Testament (e.g., in Acts) as a highly derogatory term, usually meaning fraud and charlatan. One particular figure described as a magus is Bar-Jesus, or Elymus (in translation). Saul (Paul) describes him as a Jewish false prophet:

> You son of the devil, you enemy of all righteousness, full of all deceit and villainy, will you not stop making crooked the straight paths of the Lord?[13]

This is strong stuff and seems totally unlike the good and kindly Magi of the Nativity.

In a third form of magi described in the Old Testament, the magi were experts and ambassadors who were sent as envoys to other countries to represent the royal family.

"Magi" is a somewhat neutral term, hence the debate about its meaning. The term "astrologers," however, makes strong assumptions about motivations. I therefore dislike this alternative translation, even if it may be an accurate description of their function. Greetham seems to think so, too, suggesting that for the Magi, "A knowledge of the night sky was therefore essential. Is this what our Magi were, religious, scholarly envoys? It does seem to fit better the kind of 'Magi' we find in Matthew's account. . . . They were 'searchers of wisdom.' If you must have a translation for 'Magi,' I think that 'Wise Men' is as good as any."

Where Did the Magi Come From?

It is generally assumed that the Magi came from Babylon, or at least that they came from the region of Babylonia near the city of Babylon. There is some evidence and legends, however, that suggest the Magi came from Persia, which was well to the east of Babylon and more distant from their generally supposed point of origin.

There are, as we have seen, important Persian connections with some of the characters and events described above. When I originally started to research this book I believed that the Magi were probably from Babylon, or at least from the surrounding regions. I am now far less sure and find the "Persian hypothesis" increasingly attractive. Later in this chapter we will examine different possible points of origin for the Magi in more detail, and we will see why Persia makes an interesting alternative.

The "traditional" view that the Magi were from Babylonia is attractive because the astronomical records from Babylon and the surrounding cities extend back two millennia before the birth of Jesus, and the history of these cities dates back even further. Many of the cities of the region, such as Babylon, Ur, and Ninevah, are referred to extensively in the Old Testament, revealing a cross-fertilization of ideas, traditions, and knowledge.[14] Because of the "Babylonian captivity," to which we will return later, the Babylonians would have been familiar with and steeped in Jewish traditions and prophecy, such as the messianic predictions. Because a knowledge of astronomy would be important to the Magi if they were to see and understand the Star, the notion that they came from a region of highly advanced astronomers seems both logical and sensible.

Colin Humphreys, the British researcher into the Star of Bethlehem, points out that there is also a strong tradition that places the Magi in Arabia or Mesopotamia. In A.D. 160 the early Christian writer Justin Martyr wrote of King Herod that the "Magi came from Arabia to him." The Reverend Greetham, on the other hand, says that the early Christian church believed that the Magi were Persian. To add weight to this belief, the Apocryphal Gospel of the Infancy states in its account of the Magi, in chapter 7:1, that the Magi came to Jerusalem "according to the prediction of Zoroaster," which certainly suggests that the Magi were Persian.

Let's now examine some of the science and culture of these regions to see what further clues we can obtain from them. As we know much more about Babylon than about other regional powers, we will start our search there.

Babylon and the Surrounding Regions

Four kingdoms flourished in the region at the time of Christ. One was the kingdom of Babylonia, which lay between the Tigris and the Euphrates rivers in the fertile flood plain irrigated by the two. Today this region is central Iraq. The modern-day province of Babil, south of Baghdad, contains the ruins of the ancient city of Babylon, which was built to be the capital of Babylonia. Babylon was situated approximately sixty miles from Baghdad, halfway between today's small towns of Al Hillah and Al Musayyib.

Another was the kingdom of Assyria, which occupied a region around the bifurcation of the Tigris River, around 250 miles upstream from Babylonia. Nineveh, its principal city, famous from the Old Testament book of Jonah, lies just slightly north of the modern-day city of Mostul in the Kurdish region in northern Iraq. Despite the frequent confusion between the Assyrians and the Babylonians, it was the former, the first great civilization of the world, who were among the first peoples to make and record astronomical observations on a systematic basis.

The third of the kingdoms in the region was Mesopotamia, whose name comes from two Greek words meaning "between the rivers." It covered the area between the two great rivers that cross the region, the Tigris and the Euphrates, and extended to the north bank of the latter.

The fourth kingdom, Chaldea, lay to the south and east of Babylonia, on the south bank of the Euphrates, and extended across to the Persian Gulf. Chaldea was on the border of the scorched region of the Arabian desert and Arabian peninsula. The remains of the city of Ur, its capital, lie buried some tens of miles west of the town of An Nasiryah (see fig. 2.1 for a map of the region).

The first modern civilization flourished in the city-states of Sumer in the southern part of the Babylonian kingdom well before 3000 B.C., and it included the first great cities built by mankind. Because of its location between and on the banks of the Tigris and Euphrates, this area has been a battleground for as many millennia as history has recorded, as one army after another tried to conquer and dominate these fertile lands. A fertile oasis in the middle of a desert is an inviting target for the army of a neighboring, less fortunate kingdom, and so wars have been fought over it ever since. The Sumerians were the first to invent large-scale warfare, and cities such as Ur, Kish, and Uruk honed their skills by fighting one another for supremacy of the region. The result was predictable: Sargon the Great, who had unified the people of the desert lands, looked with jealousy at the more fertile territories occupied by the warring Sumerians and realized that they were vulnerable to conquest. Sargon's army, united—if only by fear of their leader—was more than a match for the disunited Sumerians. This conquest, carried out at the start of the third millennium B.C., was the first of many invasions of the region.

Babylonian Language and Tablets

One of the greatest Babylonian contributions to science and to learning in general was the library set up during the reign of King Ashurbanipal, who lived from 668 to 626 B.C. The library, which has come to be known as the Library of Ashurbanipal, was an important archive located in the city of Ninevah. It was the main repository of Babylonian knowledge and records. Sadly, little has survived of these archives.

Above and beyond the fragmentary state of the records, archival research has been hampered by the problem of translation of the tablets. The records were written in Sumerian, an ancient precursor of Babylonian, yet it had died as a language one thousand years before the great library of Ashurbanipal was even established. Still, there was a reason for writing the archives in the completely dead Sumerian language. Doing so ensured that the contents of the tablets would be accessible only to a highly controlled elite of scribes and intellectuals. In this way, potentially dangerous knowledge was carefully limited to a few people who would not spread it. Written in the wedge-shaped characters called cuneiform, the Sumerian language is particularly difficult to decipher. It requires an advanced skill that only a select few experts have mastered, many of whom have literally dedicated their lives to the challenge.

The surviving tablets from the library have been deciphered thanks to the persistence of the translators. But it is also due to a lucky break: a few of the recovered fragments are written in Sumerian cuneiform on one face, with a Greek translation on the other. These fragments may possibly date from as late as A.D. 200. Important for us in this book is that a significant number of these tablets contain ancient Babylonian astronomical records, covering a wide range of astronomical phenomena. Many of the tablets appear to be astronomical diaries, listing phenomena in columns, one for each year.

Christopher Walker, an expert in the history of the Middle East at the British Museum in London, points out that few Babylonian tablets of any kind have ever been recovered, despite extensive excavations. The British Museum possesses about 90 percent of them, though most are tiny fragments recovered from a number of different sites. In many cases only a small part of the tablet referring to a particular year has been found.[15] Even when a number of smaller pieces have been reconstructed into a more complete tablet, the result is usually still just a small fraction of the original. None of the pieces I was shown on my visit to the British Museum were even as large as my hand.

In addition, there is a curious detail concerning these diaries. According to Walker, a number of the surviving fragments in different museums are copies of the same information. Why do these same tables of data keep appearing again and again? One reason is that the Babylonians did not want to lose such information, and so kept transcribing their records, much like Western monks did during the Dark Ages. There is direct evidence of this: the Venus Tablet describes observations made some one thousand years before the tablet was even written. The scribe was apparently preparing an astronomical diary and had either found, or had been given, these earlier observations and so incorporated them into his compilation. Middle East experts state there is much evidence that this was a systematic practice: old observations would be recycled, perhaps many times over the years. The problem is to determine the historical context of the Babylonian records we possess. Do we have really relevant and important data? Or are the tablets that have survived merely the dross that was thrown out as worthless? According to Walker, some evidence points to us having only the tablets that had, in many cases, been discarded as useless. Experts on the Babylonian tablets cannot answer these important questions, but Christopher Walker fears the worst.

Babylonian Astronomy

Although a great deal is known about the history of Babylonia from the large number of artifacts that have been excavated and have come to rest in the museums of the world, far less is known about their astronomy and science in general. We do know that they were greatly interested in the stars and planets. They also probably counted by using the sexagesimal system, in sixties and not in tens, which they used for making astronomical calculations. But little is known about their actual observations. Only a few, such as the Venus Tablet (see below) and the observation of Comet Halley in 164 B.C., are quite well known. Our main reason for this scanty knowledge is the small number of tablets that have been recovered and their generally poor state.

We know even less about the astronomers who made these records. Hardly any names have survived, and the dates when they were active are not well established. We do know that two of the greatest Babylonian astronomers were Naburiannu (around 500 B.C.) and Kidinnu (around 350 B.C.), but little is known about their work.

One of the oldest known astronomical records in the world is the observation of an eclipse of the Sun, made at Ugarit in what is now Syria. The text states that "the Sun was put to shame and went down in daytime." The date was not explicitly recorded, but it was noted that the eclipse occurred at the New Moon of the month of Hiyar, which corresponds to April/May. When the month and specific location at which the eclipse took place are known, it is a simple exercise in computing to work out the exact date of the observation. According to calculations by the British expert in ancient astronomical observations, Richard Stephenson, the eclipse can only have been the one of May 3, 1375 B.C.; no other eclipse fits.

A Spanish writer named Alberto Martos-Rubio suggests that the British Museum tablets contain the record of an eclipse of the moon seen in 2053 B.C. This eclipse, he calculates, was a partial one, with a maximum of 60 percent of the lunar disk covered by the shadow of the Earth. It would have been seen on March 28, 2053 B.C., with its maximum at 18:55 universal time. Other ancient eclipses were observed at Gibeon in Palestine on September 30, 1131 B.C., and at Nineveh in Assyria on June 15, 763 B.C.. The record of the Ninevah eclipse bears witness to the bellicose history of the region, as well as to the interest the Babylonians showed in the heavens: "Insurrection in the city of Assur. In the month of Sivan the Sun was eclipsed."

It is these records and the fact that the tablets were written in the ancient Sumerian language that leads one to believe, with a high degree of confidence, that astronomy in the region started in Sumer more than two thousand years B.C. Unfortunately, we have only a minuscule fraction of the Babylonian observations that were once set down on the clay tablets.

Despite their achievements in other fields and comparatively high level of advancement, it seems that the Babylonians were not familiar with the Saros cycle (see chapter 9) and could not, therefore, predict eclipses, at least initially. It is possible they may have gained this knowledge around the start of the sixth century B.C., conceivably through contacts with other civilizations. This is a speculative suggestion, however, and not based on any firm evidence. As the Babylonians appear to have observed far more eclipses, both of the Sun and of the Moon, than any other type of phenomenon, it is surprising they were not more familiar with the Saros cycle. The strongest statement we can currently make about their capability to predict eclipses is that some evidence suggests that the court astronomers could predict them for a particular coming month.

The Late Babylonian texts (700–50 B.C.) contain a wide variety of astronomical records. The first eclipse known to have been observed from Babylon after its foundation was on September 26, 322 B.C., long after the fall of the city to the invading Persians. The last surviving Babylonian observation that can be unequivocally dated is from A.D. 46 (although, as we have seen, it is possible that the bilingual texts may have come later). Thus we know that the Babylonians were still actively observing the heavens at least half a century after the birth of Jesus. This is at least consistent with the possibility that the Magi were Babylonians, or at least living in Babylonia.

Apart from eclipses, which seem to have been particularly important astrologically to them, the Babylonians also recorded a certain number of comets. We also find detailed observations of the planets and of occultations. No bright novas or supernovas appear to have been recorded at any time, however. This is not surprising: they are relatively rare events, so it is possible that tablets containing such observations have not yet been found. A civilization that took great interest in the sky would certainly notice such spectacular phenomena.

The peoples of the region knew of the planets and observed them in some detail. Because of the highly varied nature of the population that inhabited the Tigris-Euphrates region, a planet could have many different names in the different cities and kingdoms. Mercury, for example, was called by as many as six different names: Biib-bou for the Sumerians; Goud-oûd for the Assyrians and Chaldeans; Sekhès, Ninob, Nabou, and Nébo for the Babylonians.

Much more is known of Babylonian astronomy than of their astrology. In fact, almost all the known Babylonian records are astronomical rather than astrological. Virtually the only known astrological texts are from Nineveh and date from the seventh century B.C. and are collectively known as the "Omen texts." Of the known predictions, some deal with the astrological implications of conjunctions, showing that the people of the region were not ignorant of this type of phenomenon. Of these few conjunction-based predictions, even fewer concern what we might call "the West," termed *amurru* in the texts. As "the West" would include Palestine, there is no direct evidence to link Babylonian astrologers with Jewish messianic predictions—a worrying state of affairs if the Magi we seek were indeed Babylonians.

Babylonian observations of the planets date back many centuries. One of the oldest astronomical observations known from anywhere in the world is a Babylonian observation of Venus dating to around 1700 B.C. The Venus Tablet, as it is known, a clay tablet on which a series of comments and observations made by the Babylonians are recorded, was found at Konyunjik and taken to the British Museum. The Venus Tablet states that, when Ishtar (the Babylonian name for Venus) appears, "rains will be in the heavens," and, when it returns after an absence of three months, "hostility will be in the land, the crops will prosper." As these comments indicate, it was Ishtar who was thought by the Babylonians to be responsible for the fertility of the land. As with Mercury, first recorded in a

Chaldean inscription from 265 B.C., which was honored with such works as a stage in the Tower of Nimrod in the south of Babylonia, temples were set up to Venus in various places, including one at the city of Nineveh. The Babylonians venerated Ishtar as the mother of the gods.

Although it must have been familiar for many centuries, Mars, like Mercury, was not referenced by the Babylonians until the last centuries B.C. The name given to Mars by the Chaldeans was Nirgal or Nergal, which was the name of the Babylonian god of war, just as Mars was the Roman god of war. As in many other countries and traditions, the red color of Mars must have invoked thoughts of blood and warfare. The first recorded observation of this planet by Babylonian astronomers was its close approach to the star Beta Scorpii, which occurred in 272 B.C.

Saturn was first known to have been observed from Mesopotamia around 650 B.C., when it was recorded to have "entered the Moon," presumably a reference to an occultation. Saturn, like Mars, was regarded by Babylonian astrologers as an evil, malign planet. In contrast, Jupiter, like Venus, was regarded as a positive and benevolent one.

Many authors have suggested that the explanation of the Star was, as we will see in the next chapter, the 7 B.C. triple conjunction of Jupiter and Saturn and the subsequent planetary massing of Mars, Jupiter, and Saturn in 6 B.C. Pisces, the constellation in which both the triple conjunction and the massing occurred, is known to have been associated with the Jews in ancient astrology. This association has been proposed to explain why the Magi would have connected the conjunction with the birth of Jesus.

Given that Jupiter was regarded as benevolent and Saturn and Mars as malevolent, this triple conjunction, followed shortly afterwards by a planetary massing of the three planets, would logically have been highly charged with astrological meaning to Babylonian astronomer-astrologers. We can only speculate what they thought about a good planet meeting up with two evil ones. As we will see, though, their references to this series of events were remarkably modest.

We should also ask the same question of another planetary conjunction that has been a favorite in the past as a possible explanation of the Star of Bethlehem. This one is still favored by a small number of experts who question the current interpretation of biblical chronology. The 2 B.C. conjunction of Venus and Jupiter would have been tremendously spectacular from Babylon. The sight of two good and benevolent planets meeting and appearing to fuse into one before the astonished eyes of the observers would also have had an enormous and obvious significance. So, too, would the next night's observations, when the two planets were separate, once again, although still close together in the sky.

This 2 B.C. conjunction took place in the constellation of Leo. The Babylonian name for Leo was Ur-Ga-La, which means "the Lion" (as does the Latin name "Leo") and the constellation was represented as a lioness without a mane. In fact, all twelve Babylonian zodiacal constellations are virtually identical to the corresponding Greek ones familiar to us today.

Unlike other civilizations, the Babylonians appear not to have regarded the Zodiac as especially important, and it seems that no particular significance was given to the

Lion in Babylonian mythology. The nearest that we can get to a special meaning requires a chain of inferred reasoning. First, remember that there was a large Jewish community in Babylonia. Next, some authors believe that the star Regulus, a star in Leo, was the "lawgiver" referred to in Genesis 49:9–10,

> 9 Judah is the lion's whelp. . . .
> 10 The sceptre shall not depart from Judah, nor a lawgiver from between his feet, until Shiloh come; and unto him shall the gathering of the people be.

The argument continues that if the Babylonians were familiar with the Old Testament prophecies, they would logically have interpreted a conjunction that took place between the forelegs of Leo, close to Regulus, as an event obviously associated with the Jews.

Some experts have suggested that these conjunctions, in and of themselves, would have been sufficient to send the Magi to Palestine. This appears unlikely if the Wise Men were Babylonian: there is no evidence at all that the Babylonians had a strong interest in conjunctions. Of all the known Babylonian observations, only one refers to a conjunction between planets; it occurred on May 18, 681 B.C., between Mercury (the planet of Crown Prince Esarhaddon) and Saturn (the planet of King Sennacherib). This conjunction was interpreted by the court astrologers as implying the imminent death of the sovereign. On this occasion the predictions were found to be correct, for King Sennacherib was assassinated shortly afterwards.

One of the tablets in the British Museum is an astronomical almanac for the years 7/6 B.C. (see fig. 7.3) that covers the period of the triple conjunction.[16] It speaks

Figure 7.3. Babylonian astronomical almanac for the years 7 and 6 B.C. (The British Museum.)

explicitly of the movements of the two planets but makes neither a direct reference nor an allusion to the conjunction. The following is a translation of a small portion of the tablet:

> Month VII, the 1st of which will follow the 30th of the previous month. Jupiter and Saturn in Pisces, Venus in Scorpio, Mars in Sagittarius. On the 2nd, equinox.
>
> Month XI, . . . Jupiter and Saturn, and Mars in Pisces, Venus in Sagittarius. On the 13th Venus will reach Capricorn.

As we can see, the conjunction gave rise to no perceptible comment or interest, either in the triple conjunction (e.g., the text from Month VII reproduced above) or in the planetary massing (Month XI).

Persia and Persian Influences

Did the Magi come from Persia rather than Babylonia? This is difficult to answer largely because much less is known about the Persians and their influence in the region. But there are various interesting indications that the Magi could have been Persian in origin. One of the most important of these concerns Zoroastrianism, the religion of the Persians. This would tie in closely with Jewish messianic writings. Greetham comments that Clement of Alexandria, the second-century theologian, believed the Persian writings actually referred to the son of God. If the Magi were Persian, they would surely have watched the skies for a sign of the birth of this son.

What Was Zoroastrianism?

The Zoroastrian religion still survives today in some parts of Iran and south Asia. It was founded by Zoroaster, possibly around the year 1000 B.C., though the date is controversial; however, it certainly predates much of the Old Testament. Zoroaster is said to have been born in the town of Urmiyah, in what was historically Persia and is now modern-day Iran.

Zoroaster preached that there was one god, Ahura-Mazda ("Wise Lord"), who was the force of good in the world and who was opposed by the dark forces of evil. Thus the entire basis of the Zoroastrian religion was and is the credo "do good and hate evil," supporting Ahura-

Mazda against the universal powers of the bad, the amoral, and the corrupt.

Some Zoroastrian writings appear to be messianic. They predict that a son of Zoroaster will be born many years after his own death to a virgin who bathed in a lake in which Zoroaster's semen was preserved. The son of Zoroaster will raise the dead and crush the forces of evil. This has been interpreted by some as a prophecy of the birth of Jesus.

There are a few other direct links between the Nativity and Persia besides the connection between the Medean Magi and the Wise Men or Magi of Matthew. Roger Sinnott comments that, when Marco Polo passed through the small Persian village of Saveh, the inhabitants told him that the Magi had set out from there. This village is now a small town situated eighty miles southwest of Tehran in modern Iran. Sinnott points out this legend is not unique to Saveh, that other towns make similar claims, so we should take such stories with rather more than just a grain of salt unless there is independent evidence to support them. It is intriguing, however, to find such a widespread belief that the Magi had set out from that region; perhaps there is some truth in the legends, after all.

Another link between Persia and the Magi is the fund of paintings and carvings from the earliest centuries A.D. These were created long after the birth of Jesus and are thus an unreliable and dangerous ally. Clearly they are heavily influenced by the prejudices and biases of the time when they were made. David Hughes has pointed out, however, that these earliest carvings show that Magi wore Persian dress, wearing trousers rather than the traditional robes. There is even a legend that such a carving saved the church of the Nativity at Ravenna on the Adriatic coast of northern Italy from the ravaging Persian hordes in A.D. 614. When the invading army saw the Persian figure within the church, recognizing it as one of their own symbols, they spared the building from pillage and torching.

Greetham recounts a similar story referring to the Persian invasion of Judea in A.D. 614, which was marked by the destruction and burning of Christian churches. When the Persians came to the basilica in Bethlehem, they saw the mosaic of the Adoration of the Magi depicting them in belted tunics, full sleeves, trousers, and Phyrigian caps. To the invading troops this identified the figures as fellow Persians and, as a result, they refused to destroy the building.[17]

The suggestion that the Magi might have been Persian gives rise to new and particularly difficult problems, for there was no great tradition of Persian astronomy. In fact, we have almost no records of Persian astronomical observations and no evidence that the Persians paid any significant interest to astronomy. I asked Christopher Walker at the British Museum if there were evidence of Persian interest in astronomy and he confirmed that there is little or none. This does not rule out interest in astrology, nor that they were unfamiliar with the Jewish tradition of the coming Messiah. That said, it does make it somewhat harder to believe that the Magi were Persian. It is true that the Persians may have assimilated some of the Babylonian culture and beliefs when they conquered Mesopotamia and Babylon. So with their knowledge of Jewish and Zoroastrian prophecies, it is by no means impossible that Persian Magi could have traveled a distance almost double that previously supposed.

Saveh, for example, is some 430 miles northeast of the historical site of Babylon.

What Other Options Exist?

One other possibility for the Magi's homeland is Arabia. This is based mainly on the writings of Justin Martyr (see above) and of Clement of Rome, who was allegedly killed by being tied to an anchor and tossed into the Black Sea. In his First Epistle to the Corinthians, written in A.D. 96, Clement says he associated the Magi with "the districts near Arabia." The only other support for this view is the prophecy of Isaiah 60:6, which names the gifts that would later be presented to the infant Jesus. There appear to be few other direct links with Arabia to justify our search for the Magi there, especially as we have candidates elsewhere who are much more plausible.

Where, Then, Did the Magi Come From?

The scant evidence we have, most of it indirect and much of it dating from centuries after the birth of Jesus, suggests the Magi may have been Persian, but the existence of a civilization and of astronomical records lasting over millennia make Babylonia seem a more plausible alternative. Persian interest in astronomy was marginal at best, making it more unlikely that they would be attracted by the Star. Persia, however, had a messianic tradition, so Persians had a motive that would sufficiently explain their interest in the Nativity. Furthermore, it is possible that the influence of Babylonian astronomy could have extended there, especially because we know that there was an exodus of astronomers from Babylon in pre-Christian times. If there existed a significant Jewish or, possibly, Babylonian colony in Persia, the Persians would certainly be expected to have known of their prophecies, which would have confirmed, or at least been in harmony with, their own Zoroastrian beliefs.

What about the case for Babylon? Colin Humphreys mentions a series of other important and interesting details that tie in the Magi with Babylonian and Jewish traditions. Around 586 B.C., with the Assyrian/Babylonian empire at its zenith, the Babylonians invaded and sacked Jerusalem. From then on, Babylon had a significant Jewish population, stemming from the thousands of families carried off from Jerusalem as the spoils of war. This was referred to as the "Babylonian captivity" and was the second large-scale deportation of Jews from the region, following on the sack of Samaria a century and a half earlier. Both conquests led to large numbers of Jews being displaced to the east. Although the occupation of Israel did not last long, its consequences would have been far more enduring.

This Jewish presence could certainly have led to a strong cross-fertilization of Jewish prophecy with Babylonian science. Babylonian astronomers and astrologers may thus have become familiar with Jewish prophecies of the coming Messiah, making it reasonable for the Babylonians to react to the appearance of a star in the sky announcing the birth of the new king. It is even possible that the Magi were themselves descendants of the original diaspora Jews. Thus the Babylonians had the means, the opportunity, and a possible motive to identify a particular phenomenon with the Messiah. The Persians had a clear motive in the prophecies of Zoroaster, but they were not likely to have been influenced by astronomical observations.

If I were forced to decide between these two options as the Magi's homeland, I would select Persia. But one must also heed Christopher Walker's words: "If the Magi ever existed, I think that the only plausible explanation is that they were Diaspora Jews." Not all authorities agree with this possibility, and some disagree violently. If the Magi were Jewish exiles from Babylon, some important questions would be answered regarding the Magi's interest in the Star and in the birth of the Messiah. Alas, this conclusion begs a major question: if the Star of Bethlehem were really that important and that obvious to the exiled Jews, how is it that King Herod and his men seemingly knew nothing of it?

8

Triple Conjunctions: A Key to Unlocking the Mystery?

WHETHER THEY came from Babylonia, or from Persia, the Magi were clearly steeped in astronomy, astrology, religion, and, especially, prophecy. They would be moved by certain astronomical phenomena and would find great meaning in them depending, for example, on the constellation in which the phenomenon occurred; they might well ignore other phenomena or assign lesser importance to them. For the birth of a new king, one who would fulfill the prophecies, a special sign would have been required. The Magi had seen comets come and go; they were aware of conjunctions and occultations through the centuries; of meteors, planet massings, novas and supernovas, and the like. Each of these events could have provided an extraordinary visual experience for them and offered astrological meaning as well, but apparently none of them heralded a king's birth. For whatever reason, the pieces did not fit for the Magi. And, because we know the approximate date of the Nativity, we also know that many of these heavenly phenomena were way off the calendar. If we are to believe the account given in Matthew's Gospel, even though it was written down many years after the Nativity it describes, then we must believe that there was a Star and the Magi followed it. We must also consider, however, that perhaps the Magi were spurred not *only* by the Star. What if there had been a *series* of celestial events, a series so compelling that the Magi fiercely persisted to observe the heavens for a further hopeful sign, then another, and then, finally, the crowning achievement: they saw the long-awaited Star that had been prophesied. If this was true, what might have been the first event?

One of the events may have been the triple conjunction of Jupiter and Saturn in 7 B.C. We saw earlier that a conjunction occurs when one body (a planet or the Moon) passes either north or south of another one in the sky. Conjunctions could involve two planets, a planet and a star, a planet and the Sun, or even the Moon and a star. A triple conjunction is a little different and an extremely rare event involving a particularly intricate set of movements of two planets. Instead of one planet making a single pass close to another in the sky, the two bodies pass, separate, pass a second time, separate again, and then pass a third time before separating for good. Such a triple conjunction can happen only with the superior planets—the ones outside the Earth's orbit—and is best seen in the case of Jupiter and Saturn.

Any pair of exterior planets can give rise to a triple conjunction, but these are relatively more common for the most distant planets.[1] For example, Mars is in conjunction with either Jupiter or Saturn approximately every two years. Between the years 1800 and 2000, for example, there were eighty-nine normal conjunctions of Mars and Jupiter but only two triple conjunctions, one in 1835–36 and another in 1979–80. For Saturn, the statistics are even more heavily loaded: ninety-nine normal conjunctions, but just one triple conjunction with Mars. In contrast, Jupiter and Saturn have had eighty-nine normal conjunctions and eleven triple conjunctions in the last two thousand years (11 percent triples). Uranus and Neptune, even though they reach conjunction only every 176 years, actually gave rise to far more triple conjunctions (63 percent) than normal ones (37 percent) over the two millennia. Overall, triple conjunctions between any pair of planets happen, on average, around every two centuries. This is strictly an average, however, and wide variations can occur, as we will see. However, an astronomer or astrologer would not normally see a single triple conjunction in his or her whole lifetime.

Let's take a closer look at the triple conjunctions of Jupiter and Saturn, which interest us most because of their relevance to the story of the Star of Bethlehem. The three individual encounters, or conjunctions, in the sky of the two planets are usually spread over some seven months. During this time, other planets may add themselves temporarily to the grouping, giving rise to what is called a "planetary massing." During a planetary massing several planets may all appear to be quite close together, although they may be separated by 5 or even 10 degrees, and hundreds of millions of miles.

A particularly spectacular planetary massing occurred several years ago, with Mars, Venus, and Jupiter all close together in the evening sky. In mid-June 1991, the three planets were in Cancer and visible for several hours after sunset. On June 16 they were within a circle of diameter of 2 degrees; the crescent Moon was just below, and, as the sky darkened, they made a stunning group, sinking slowly in the west. Even after the Moon had left the arrangement, the three planets were close together for several more nights. I was fortunate enough to see all of this develop while observing at Teide Observatory in Tenerife, and I can vouch for the tremendous impact it had on the witnesses of this gorgeous and much-photographed event.

The difference in brightness among the three planets during this grouping was enormous: Venus was at magnitude −4.3; Jupiter at magnitude −1.8; and Mars was at magnitude +1.7, so that Venus was about 250 times as bright as Mars. This contrast among the three planets did not make the massing any less spectacular. On June 18, Venus passed 69 minutes north of Jupiter; five days later, on the 23rd, Mars passed 13 minutes south of Venus and, at the same time, Jupiter moved slowly away from the pairing. Three weeks passed before there was another striking grouping, when Venus and Mars, still very close together in the sky, passed both sides of the star Regulus.

Because such a planetary grouping causes great interest even today, one can see that such an event could have been even more important to the Magi, with their interest in the planets and planetary movements and alignments. It is not surprising that they have created such interest over the years and have often been candidates for the Star of Bethlehem. In fact, it has been almost four centuries since a conjunction was first mentioned in the context of the Christmas Star.

A triple conjunction visible to the naked eye is a phenomenon exclusive to Mars, Jupiter, and Saturn; Uranus and Neptune can take part in triple conjunctions, but these can be followed only photographically or with binoculars. Usually the two planets involved do not pass particularly close together. Only rarely is their separation much less than a degree (twice the diameter of the Moon). For this reason, triple conjunctions are usually not as spectacular visually as planetary massings, or even some normal conjunctions, but they could have had a particular astrological significance to the Magi because of their rarity and their uniqueness among planetary phenomena.

A conjunction of Jupiter and Saturn will occur on average every 19.86 years. Jupiter, which takes 11.79 years to go around the Sun once, makes one complete orbit and then nearly two-thirds of a second before it "catches up" with Saturn, which has a 29.46-year orbit around the Sun. Each time Jupiter catches up, the two planets line up briefly and appear close to each other in the sky for a few days.

Usually, Jupiter charges past in its swifter smaller orbit, like an athlete in the inside lane of a racetrack, passing rapidly by its slower companion, Saturn, and then receding into the distance. In such circumstances there is just a single moment when the two planets line up: a single pass and a single conjunction. Occasionally, the parallax, or the changing viewpoint caused by our own Earth's movement around the Sun, added to the movement of Jupiter and Saturn, causes us to see a triple encounter between the two planets instead of a single one. For this to happen, the Earth, Jupiter, and Saturn all have to be positioned in their orbits quite precisely, so that the Earth's movement from one side of its orbit to the other, over six months, makes the two outer planets line up not once but three times. This can happen only if the Earth "overtakes" both Jupiter and Saturn at just the same moment when Jupiter overtakes Saturn.

A rule of Jupiter-Saturn conjunctions is that a triple conjunction is always followed by a normal, single conjunction twenty years later. Usually, a series of normal conjunctions follow, each twenty years apart, until the next triple conjunction happens. Occasionally, a second triple conjunction may occur only forty years after the previous one: this happened for the two most recent triple conjunctions of Jupiter and Saturn. In 1940–41, most of the world's astronomers, apart from the Americans, had other things on their minds besides the behavior of the planets: the vast majority of them were thinking more about survival in the world war that was increasing ever more in intensity than about the abstract problems of astronomy. However, a triple conjunction took place in August and October 1940 and February 1941. A new triple followed in 1980–81, when the two planets encountered each other on New Year's Eve 1980 and, later, in March and July 1981.

Whenever two triple conjunctions are separated by just forty years, like the one in this century, the following one will not occur for several hundred years. Anyone who missed the 1980–81 triple conjunction will have to wait until 2238–39 and then 2279 for the next one. In extreme cases the interval may be more than four centuries, even though normal conjunctions will keep occurring every twenty years, as regular as clockwork, in the meantime.

The first mention of a triple conjunction in the context of the Star of Bethlehem was made by Johannes Kepler, who, as we saw a couple of chapters ago, observed the supernova of 1604. A few months earlier, in December 1603, he saw a conjunction of Jupiter and Saturn followed by a massing of Mars, Jupiter, and Saturn. Early in 1604, Kepler calculated back in time and determined that a similar conjunction followed by a massing with Mars had also been seen in 7 B.C. He did not suggest that this was the explanation of the Star of Bethlehem. Kepler, as court astrologer to the Holy Roman emperor, Rudolf II, was undoubtedly fascinated by the configuration; but he was even more impressed by the supernova that followed, close to the three planets, in the constellation of Ophiuchus. Kepler, in fact, believed that the Star of Bethlehem was a nova similar to the one he had just observed. Despite this, most discussions of the Star incorrectly attribute the triple conjunction theory to Kepler. In fact, the first serious suggestion that the triple conjunction was the Star of Bethlehem was not made until 1825. Apparently, a German astronomer and philologist named Christian Ludwig Ideler had misunderstood Kepler's comments and promoted this theory. It is only comparatively recently, during the last two decades, that the triple conjunction theory has become popular.

For a triple conjunction to be considered a plausible candidate, we again have to be sure that they are neither too common nor too repetitive. How frequent were triple conjunctions of Jupiter and Saturn in biblical times? Too many in too short a space of time would be a severe blow to the theory's credibility.

A search for all conjunctions of the two planets, in which the two passed within 3 degrees of each other in a north-south line, led to many encounters. Only rarely is the separation between the two planets as much as a degree and a half during a conjunction, and it is never more than 2 degrees; in this way, by seeking separations less than 3 degrees, we won't miss any—it pays to be conservative in our search criterion. Similarly, few of the conjunctions are particularly spectacular: out of 122 conjunctions of Jupiter and

Saturn in the first two millennia A.D., only seven led to a minimum separation of less than 10 arcminutes.

In total, there were no fewer than sixty-four conjunctions of Jupiter and Saturn between 1000 B.C. and A.D. 1, some of them close ones and some involving other conjunctions with other planets. A reasonably close conjunction of Jupiter and Saturn (45-minute separation) occurred in late April of 126 B.C. and was preceded by massings of Mercury, Jupiter, and Saturn in mid-March, and of Jupiter, Venus, and Saturn in early April. This latter massing gave rise to an extremely spectacular conjunction between Venus and Jupiter.

Of the sixty-four conjunctions of Jupiter and Saturn between 1000 B.C. and A.D. 1, there are only seven sets of triple conjunctions, listed in table 8.1, which gives the years, separation, and constellation in which the encoun-

TABLE 8.1 Triple Conjunctions Observed between 1000 B.C. and A.D. 1

Years	*Minimum Separation*	*Constellation*
980–979 B.C.	38 minutes	Pisces
861–860 B.C.	55 minutes	Pisces
821–820 B.C.	22 minutes	Leo
563–562 B.C.	68 minutes	Taurus
523–522 B.C.	65 minutes	Virgo
146–145 B.C.	10 minutes	Cancer
7 B.C.	58 minutes	Pisces

NOTE: Note that two triple conjunctions far more spectacular than that of 7 B.C. happened in the second and ninth centuries B.C. A much better triple conjunction in Pisces than the 7 B.C. event also occurred in the tenth century B.C.

ter occurred. The table shows that triple conjunctions are quite rare, but not excessively so. There is a wide range of different types of triple conjunction: some are quite spectacular, with the two planets separated by very little distance in the sky; others are far less noteworthy to the observer.

Table 8.1 is revealing in another important sense. We can see that a spectacular triple conjunction of Jupiter and Saturn, the best of the first millennium B.C., occurred in 146–145 B.C. This is interesting because it happened less than a century and a half before the much commented upon and studied 7 B.C. conjunction, which is a popular choice as the Star. If the only factor involved were the conjunctions alone, the Magi, on seeing this spectacular triple conjunction, would have arrived in Jerusalem 139 years early! While the 7 B.C. conjunction would not have been that striking to the observer even at its closest approach, the 146–145 B.C. event was completely different. In each of the three individual conjunctions in 146–145 B.C. the two planets were separated by less than half the diameter of the Moon, and at closest approach were only a third of its diameter apart.

There is another important factor to consider. Where did the conjunction occur? The Magi were probably astrologers more than astronomers, and the constellation in which the conjunction occurred would have conveyed meaning to them. For example, a triple conjunction in Pisces, the constellation associated with the Jews, would be an important sign. We can see from table 8.1 that there were actually three triple conjunctions in Pisces, though two of them happened almost a millennium before the 7 B.C. event. It could be argued that these earlier ones took place before the biblical prophecy of the coming of the

Messiah had been made, which means they could hardly have been relevant to the Magi. Both of these ancient triple conjunctions antedate the construction of Babylon, which occurred in the seventh century B.C. But they certainly did not antedate Sumerian and Assyrian astronomy, so could certainly have been observed by the forefathers of the Magi. Should these older triple conjunctions be ignored simply because they occurred too early, so that the Magi were just not ready for them? Just when did the Magi and their ancestors begin their long vigil?

This final question is a thorny point and difficult to answer. The writings of the Old Testament go back many centuries B.C. and some of the events reported probably go back several millennia. The coming of the Messiah was probably prophesied several centuries before the founding of Babylon, but we may never know exactly when it was done. For safety's sake, we should treat any event observed after 500 B.C. as though it could conceivably have been "mistaken" for the Star, and possibly we should even include events that took place several hundred years earlier, too. The former date corresponds approximately to the date when a large part of the Jewish population of Jerusalem was dispersed by the Babylonian conquest of the city, and thousands of Jewish people were transported to Babylon. From this point onward there was a strong Jewish influence in Babylonian astronomy and, through this, a cultural diffusion of Jewish traditions and prophecies throughout other kingdoms of the region.

What exactly is the cut-off date before which we need not search? Here there are many doubts: the oracle of Balaam mentions the Star (see chapter 1) and was almost certainly written before the fifth century B.C., possibly as early as the eighth century B.C. In other words, any event later than 500 B.C., and possibly an event as early as 800 B.C., *might* potentially have been taken for the Star of Bethlehem.

Perhaps the single most important factor is not the fact that the 7 B.C. triple conjunction occurred in Pisces. It is just as possible, as has indeed been suggested by some authorities, that the key factor in the Magi's identification of the Star of Bethlehem was the combination of a triple conjunction with other events. This identification could even have been a combination of several events totally unrelated to the triple conjunction. This means that it is important, if we want to understand the Star, to look at these historical triple conjunctions to see if the 7 B.C. event was unique in some way that the Magi would have noted, or whether other triple conjunctions were associated with phenomena similar to the 7 B.C. event. If we find that the 7 B.C. conjunction was indeed unique, then its candidacy as the Star of Bethlehem, or perhaps its precursor, becomes much stronger and more plausible. If not, it is much less likely that a triple conjunction could, on its own, have much significance. Let's look briefly here at each of the triple conjunctions of the first millennium B.C.

The 989–979 B.C. Triple Conjunction

This first triple conjunction of the millennium began with its first encounter on May 5, 980 B.C., followed by a conjunction on September 25, 980 B.C., and March 11, 979 B.C. The separation was 42 minutes for the first and third conjunctions and 38 minutes for the second, hence at closest the two planets were separated by little more than the width of a Full Moon's diameter. During this triple conjunction the Moon passed extremely close to the two planets many times, occulting both Jupiter and Saturn on numerous occasions. Of course, the fact that an occultation occurs, as seen from somewhere on the Earth, does not mean that it would necessarily have been seen either from Babylon or from Jerusalem.

In fact, this is exactly what did happen: most of the occultations, except for one, would have been unobservable from Jerusalem and Babylon. During the late evening of August 26, 980 B.C., the Full Moon would have occulted Jupiter, which was high in the southern sky. Although the glare from the Moon would have been particularly strong, it would still have been a quite spectacular occultation to observe visually, particularly in view of the fact that Jupiter and Saturn were closing for their second conjunction at the time. Astrologically, the occultation would have been full of meaning for two reasons: first, because a royal planet, Jupiter, was "killed" by the Moon and then reborn; second, because this conjunction took place in Pisces, which, as we have mentioned, would hold importance to the Jews. These are exactly the types of associations we are searching for when looking for the Star.

The 861–860 B.C. Triple Conjunction

This conjunction also occurred in the constellation of Pisces. The circumstances were similar to the 7 B.C. triple conjunction in that the separations of the two planets (57 minutes, or slightly less than one degree and twice the diameter of the Full Moon, on July 3, 55 minutes on August 8, and 63 minutes on January 1, 860 B.C.) were almost identical to those seen in 7 B.C. But, unlike the 7 B.C. conjunction, no planetary massing was associated with this event. Jupiter and Saturn were rather isolated from other planets in the sky until well after the triple conjunction was over. However, as in the previous triple conjunction, there was a prominent occultation of Jupiter by the Moon. At 21:05 local time on October 25, 861 B.C., with Jupiter 46 degrees high in the southeastern sky, the large gibbous moon, just a few days before full, occulted the planet as seen from Babylon. This was one of a long series of occultations of Jupiter and Saturn by the Moon that took place at around this time, although only one other one was even theoretically visible from Babylon. That second occultation happened in full daylight during the early afternoon of December 19, 861 B.C., and hence could not have been observed with the naked eye.

The 821–820 B.C. Triple Conjunction

On this occasion, the conjunction occurred in Leo, below the lion's hindquarters. This was a particularly fine triple conjunction, with the two planets passing each other within less than the diameter of the Full Moon on all three occasions. On November 16, 821 B.C., their separation was 25 minutes. By January 7, 820 B.C., this reduced to just 22 minutes, increasing again to 28 minutes at the third encounter on June 5, 820 B.C. As the two planets separated in the sky and came closer to the Sun, a prominent planetary massing occurred. Four planets—Jupiter, Saturn, Mars, and Mercury—all fell with a few degrees of one another around

July 21, 820 B.C. This massing would have been obvious and quite spectacular, low in the west, in the evening twilight sky.[2]

The 563–562 B.C. Triple Conjunction

This was the least spectacular of the seven triple conjunctions of the first millennium B.C. It took place between the Hyades and the Pleiades star clusters in Taurus. The smallest separation between the two planets was 68 minutes on July 27, 563 B.C., with the second conjunction on November 1 at 75 minutes and the third, on February 10, 562 B.C., at 76 minutes. While no massing of planets occurred, there was a comparatively important, although not outstanding, conjunction of Mars and Jupiter on June 5, 563 B.C. (separation 26 minutes), which preceded the triple conjunction.

The 523–522 B.C. Triple Conjunction

This too was a comparatively unexciting event, with the two planets always more than a degree apart. It was, however, marked by a spectacular planetary massing and a further conjunction, making this a rather important triple conjunction, if only by association. The three conjunctions between Jupiter and Saturn occurred on December 16, 523 B.C. (65 minutes separation), March 14, 522 B.C. (65 minutes) and July 10, 522 B.C. (71 minutes). Around August 22, 522 B.C., a few weeks after the final encounter between Jupiter and Saturn, an important massing of Jupiter, Saturn, and Venus occurred. This was followed, three days later, by a spectacular conjunction of Jupiter and Venus when, although their separation was slightly more than a degree (69 minutes), the brightness of the two bodies would have made their approach to each other particularly beautiful in the evening sky.

The 146–145 B.C. Triple Conjunction

We should pause here for a moment, if only to wish to have experienced this triple conjunction, the finest one of the first millennium B.C. It was not associated with other phenomena, planetary massings, occultations, etc., but it would have been a spectacular event to observe. The conjunctions took place in the constellation of Cancer between October 146 B.C. and May 145 B.C. Cancer is a quite faint constellation: it lies between the much more spectacular Gemini (to its west) and Leo (to its east). The first close approach of Jupiter and Saturn was on October 18, 146 B.C., with the two planets separated by 11 minutes (a third of the diameter of the Moon), the second, eight weeks later, on December 10, was slightly more distant (15 minutes, or half the diameter of the Full Moon), and the third and closest approach (10 minutes separation) took place on May 4, 145 B.C. The planets did not get quite close enough together to fuse into one, but it must still have been a spectacular event to observe!

Our scrutiny of previous triple conjunctions and our comparisons with the one of 7 B.C. lead us to conclude once more that the 7 B.C. event was "just another triple conjunction." There was nothing particularly outstanding about it, either in appearance or circumstances, to make it stand out from the crowd. Although one can point to the later planetary massing in 6 B.C., other triple conjunctions were associated with better and more spectacular massings, close conjunctions, and even occultations. In other words, the 7 B.C. triple conjunction was, unless we are missing something important, nothing special.

The 980–979 B.C. and 861–860 B.C. triple conjunctions both occurred in Pisces, as did the 7 B.C. event. The first was a much better conjunction than that of 7 B.C., while the second was similar to the 7 B.C. conjunction. Based only on the quality of the conjunction itself, we would expect the first of the three to have been the Magi's star. The main difference between the first two and the one of 7 B.C. is the nature of the associated event: in 7 B.C. there was a later massing of Jupiter, Saturn, and Mars; in both 980 and 861 B.C. there was an occultation of Jupiter by the Moon that was visible from Babylon. As has been suggested elsewhere, the occultation would surely have had stronger, more positive connotations than the massing. With the disappearance of Jupiter and its later rebirth in Pisces, the idea that the event marked some kind of important, perhaps even royal birth in Palestine would surely have occurred strongly to the Babylonian astronomers. The fact that they recorded occultations with more apparent enthusiasm than they did conjunctions only serves to reinforce this point.

In contrast, the 6 B.C. massing of the planets could have had only a negative connotation: Mars is the god of war, and the planet Mars has been associated with blood and warfare by most races. After the triple conjunction in Pisces, which promised important news from Palestine, the advent of Mars in the same constellation would seem to suggest warfare or bloodshed more than it would a royal birth.

Some authorities have associated the constellation of

Leo with the Jews and have suggested that the star Regulus may have been associated with the coming of the Messiah. In this case, the 821–820 B.C. triple conjunction would have been charged with symbolism for those familiar with biblical prophecy. It was not just a particularly close conjunction, with the two planets never approaching each other much closer than twice the diameter of the Full Moon, but, as it occurred in Leo below the lion's feet, one could argue that it would have a particularly strong symbolic significance.

In addition, this conjunction was followed by a massing not just of Jupiter, Saturn, and Mars, as in 7 B.C., but of Mercury as well. On July 21, 820 B.C., the four planets, massed together, would have set about an hour and a half after the Sun as seen from Babylon. Although they would not be visible in a completely dark sky (Mercury is almost never visible in a completely dark sky, anyway), the massing would have been perfectly observable from Babylon or Jerusalem and far more spectacular than its 6 B.C. counterpart.

In other words, the 821–820 B.C. conjunction was, in its way, more spectacular and had astrological implications that were as strong, if not stronger, than the one in 7 B.C. So why didn't the Magi take these events to be the announcement of the birth of Jesus? It seems that either (1) the theory of the triple conjunction is wrong, or (2) the triple conjunction preceded the time when the Magi started to watch for a portent of the Messiah's birth. The latter would be correct if the Magi's vigil started with the Babylonian conquest of Jerusalem, assuming that this event married Babylonian astronomy and the biblical prophecies regarding the birth of the Messiah. It is difficult to date first messianic predictions, but they may go back as far as the eighth century B.C. or slightly earlier. In this case, the 821–820 B.C. triple conjunction may have occurred slightly too early to be regarded as a messianic sign. Similarly, what are the chances of any biblical prophecy of this magnitude being fulfilled within a very few years of its making?

We can pass over the 563–562 B.C. triple conjunction, which was neither spectacular nor associated with any other special circumstance. The 523–522 B.C. triple conjunction, on the other hand, may not have been particularly spectacular but it was associated, as we have seen, with other, important events. The massing of planets that occurred in 522 B.C., after the triple conjunction, was far more spectacular than the rather poor 6 B.C. massing because Venus was more than five magnitudes (a factor of one hundred times) brighter than Mars. Although Virgo would not have had a strong astrological connection with the Jewish people, the massing, immediately followed by the conjunction of Venus and Jupiter—the two brightest planets and also the two which the Babylonians regarded as positive and benign—may have held some kind of special significance for the Magi. That said, even though the Magi may have been greatly impressed by this sequence of events, we have no reason to suspect that they would have made the association with the coming of the Messiah.

Finally, the 146–145 B.C. conjunction would also have been tremendously spectacular because of the closeness of the approach between the two planets. But it would not hold any other special astrological significance. There is no reason to associate the constellation of Cancer with the Jews, nor were there any occultations or other events of particular interest around this triple conjunction.

Thus, of the seven triple conjunctions seen during the first millennium B.C., four could potentially have had some kind of association with the Jews and could have been taken as a sign of the coming Messiah. One of these triple conjunctions was particularly outstanding and occurred in Pisces, associated with the Jews. Even one of our "less important" triple conjunctions was much more spectacular visually than the 7 B.C. conjunction. From this perspective, the only strong reason anyone could accept the 7 B.C. conjunction as the Star of Bethlehem, above these rival candidates, is hindsight: unknown to the Magi, this conjunction really did occur close to the date of the Nativity.

Some writers have been concerned that a special normal conjunction of Jupiter and Saturn might still be more significant than the 7 B.C. triple conjunction. A case in point, often discussed together with the 7 B.C. triple conjunction, is the remarkable series of spectacular conjunctions and massings of planets observed around the 126 B.C. conjunction of Jupiter and Saturn (see table 8.2). This event, these people argue, also took place in Pisces and

TABLE 8.2 The Circumstances of the Events Surrounding the 126 B.C. Conjunction of Jupiter and Saturn

Date	*Planets*	*Separation*	*Notes*
January 25, 126 B.C.	Mercury, Jupiter	16 minutes	Conjunction
March 16, 126 B.C.	Mercury, Jupiter, Saturn		Massing
April 4, 126 B.C.	Venus, Jupiter	12 minutes	Conjunction
April 4, 126 B.C.	Venus, Jupiter, Saturn		Massing
April 24, 126 B.C.	Jupiter, Saturn	45 minutes	Conjunction
April 24, 126 B.C.	Mercury, Venus, Jupiter, Saturn, Moon		Massing

would have been at least as significant to the Magi as the triple conjunction in 7 B.C. Although we have seen that an ordinary conjunction is not a very believable explanation of the Star, the events in 126 B.C. were far from ordinary. Why did some Magi not go in search of Jesus in 126 B.C.? Perhaps it was because they were waiting for something more than just an individual event.

We are still at the point where we have discovered some rather extraordinary celestial phenomena, some of which, on their own, could have been candidates for the Star, but the pieces do not quite add up. It is possible that Magi had looked very closely at triple conjunctions all along the way, and they probably regarded each of them as a sign. Those that occurred in Pisces or Leo might even have held special astrological import to them. But in the end, no one particular event would have set them off on their journey—no event screamed out "I am the sign that you have been awaiting" more clearly than any other event. Indeed, nothing in any document tells us there were any false starts.

Let's step back a bit from our world and our culture. Today we live in a world that moves at breakneck speed. Things happen fast. More and more, we expect immediate results. This was not the case in the world of the Magi. Their nature and expectations of events were different from ours. Perhaps today we would seek the one big flash that says that *this* was the Star. As the most patient of skywatchers, the Magi might have taken the view that there would be a series of signs, that maybe the prophecy would be fulfilled not by a single big flash, but gradually, sign by sign. Again, with the advantage of hindsight, I think that we can say that the Magi did find their first sign in the triple conjunction of 7 B.C. They did not yet start to prepare their camels, but it did encourage them to watch the sky even more closely for a further sign. Apart from the writings of Matthew, and perhaps in other as yet unfound documents and histories, what the Magi saw next is not recorded—at least not by the world in easy reach of the rivers Tigris and Euphrates. We must travel far to the East to discover some secrets that have been there all the time.

Is the Answer Written in Chinese?

IT WAS NOT until the 1950s that astronomers began to realize they had entirely ignored one of the most important and complete archives of astronomical data in the world. Only then did attention begin to be paid to the literally thousands of Chinese astronomical records that were known to exist and which covered not just centuries but millennia of observations of eclipses, comets, novas, supenovas, auroras, sunspots, and anything else that can be seen in the sky by diligent observers who used nothing more than the naked eye.

Though the first usage of the Chinese archives dates back some forty years, only in the past two decades have ancient Chinese and, to a lesser extent, Korean, Japanese, and Arabic records been recognized as an incredibly rich source of knowledge and information. Now, the Far Eastern archives are a starting point for many studies of astronomical phenomena, from the behavior of the sunspot cycle over many centuries to the frequency of supernova occurrence in our Galaxy.

The value of these records lies in the systematic way the Chinese astronomers watched the skies for thousands of years and recorded everything they saw that was worthy of note. They understood, or thought they understood, some of these phenomena: eclipses, comets, novas, and supenovas. They were less familiar with others: sunspots, auroras, and meteor showers. All are faithfully recorded, often in great detail, and so can be identified and studied even if the astronomers who made the observations did not know what they were seeing.

Once the true value of the Chinese archives was recognized, observations from other countries were studied and analyzed in detail as well. In many cases, we now know that Korean records, albeit they are less complete, confirm the Chinese chronicles; more recently, Arab and Japanese records have become available that help to complete the historical record and permit a further confirmation of the Chinese observations. The Arab records are only just starting to be examined in the detail they merit, and comprise one of the last great frontiers of historical research.

Sometimes Chinese and other Asian astronomers made their observations of the sky because of a simple search for meaning: like astronomers/astrologers from other cultures, they believed that many of the phenomena they saw and catalogued were auguries that would help them predict the future. In other cases, however, it seems the Chinese recorded what they saw out of simple fascination for anything new or unusual that could be seen in the heavens, whatever it may have been. In one particular way the motivation for making careful observations was urgent and crucial to them: the Chinese were terrified of eclipses of the Sun. It was tremendously important to predict them accurately.

The Chinese knew that eclipses are cyclic. What we now call the Saros cycle governs the appearance of eclipses in any one place. "Saros" is a word of Babylonian origin first used in modern astronomy by Sir Edmond Halley in 1691. When you see an eclipse, for example in Paris on any particular day, you know that another eclipse will often be visible from the same place one complete Saros cycle, or 18 years and 11.3 days, later.[1] The reason for this cycle is simple, although the Chinese who used it did not know it. Every 18 years and 11.3 days, the Sun, the Moon, and, most importantly, the node of the Moon's orbit (the point where it crosses the plane of the Earth's orbit) return to just about the same places relative to one another. Because the realignment is not exact, the eclipses shift slowly, either from north to south or from south to north, from eclipse to eclipse. A Saros series lasts for some seventy eclipses, and hence about twelve hundred years, so the Saros cycle allows the prediction of eclipses over many years, with great exactitude, but needs no mathematical knowledge at all apart from an ability to count the days exactly.

The Chinese were terrified of eclipses because they did not understand what was really happening—they did not know that an eclipse was simply the Moon passing in front of the Sun. They believed that the disk of the Sun disappeared because a great dragon was eating it. Understandably, this was a traumatic experience for them and caused great terror among the general public. Therefore one function of court astronomers was to predict eclipses, but not for astronomical reasons. Forewarned that an eclipse was imminent, they could organize the people to make a great noise at the critical time—banging drums and exploding fireworks—and thereby scare the dragon away. This technique never failed: the Sun was always restored to its former glory sooner or later!

The first eclipse ever recorded by the Chinese was one seen in either 1948 B.C. or 2165 B.C. (the exact date is uncertain). Supposedly, this eclipse led to the execution of the two court astronomers, Hsi and Ho, who failed to predict it and so did not organize any attempts to scare away the dragon. The Earth was lucky on this occasion because the Sun was still able to escape the clutches of the dragon, but Hsi and Ho did not share in this luck. They were beheaded for their unforgivable dereliction of duty. The story is probably apocryphal, but, as it is based on a reconstruction of records lost in the third century B.C. and not reconstructed until the fourth century A.D., we will never know for sure. This story does, however, illustrate graphically the importance that the prediction of eclipses held for the Chinese. If summary execution for failure seemed a more than reasonable punishment for failure to predict one, the Chinese really did take them to heart.

This first eclipse, as were all of the oldest Chinese observations, was recorded on oracle bones. Oracle bones were tortoise shells, buffalo bones, or the bones of other large animals on which the observations were carved. These bones were then buried in the earth so that they would never serve a lesser purpose. Obviously, the amount of information that can be carved onto a bone is rather limited. The oracle bones recorded eclipses more than other types of observations.

Later, the Chinese recorded their observations on silks and in books. Doing so permitted the inclusion of more details. Also, silks and books were more enduring media which, for us, can now reveal their astronomical content. It was not unusual for the Chinese astronomers to include diagrams, sketches, and drawings to illustrate their observations. Some twenty years ago, a dramatic discovery was made during an excavation at Mawangdui, where the most important of the finds was a book in which comets were cataloged and painted on silk pages. Each type of comet—twenty-nine in all—was sketched and carried a caption briefly detailing what the comet foretold. This book can be dated to the third or fourth century B.C., but it is evident that its compilation started many centuries earlier.

As we know, the Arabs, too, have a long and distinguished history of astronomical observations and chronicling. In fact, of the four known eclipses before 1000 B.C., two were recorded from Babylonia by Arab astronomers. Because Europe was submerged in the Dark Ages from about A.D. 476 to 1000, few European written records of any kind survive from this period, but both Chinese and Arab astronomers watched the skies and cataloged everything they saw during this time. Some of them made discoveries that were not duplicated in Europe until many centuries later.

Another pioneering event, made many centuries before the invention of the telescope, was the naked-eye observation of sunspots by Chinese astronomers. Even today, many people do not realize that sunspots can be observed with the naked eye when the Sun is dimmed by cloud or haze. Chinese chronicles show many naked-eye observations of sunspots over the millennia, although the observers did not realize they were really seeing a feature on the Sun's surface. These records have allowed the sunspot cycle to be investigated far back in time, well before the moment when Galileo supposedly "discovered" sunspots in the sixteenth century. One of the most surprising results has been the discovery of periods when sunspots "vanished" from the Sun's disk for many years, the only known absences prior to the famous Maunder Minimum of 1645–1715 (see p. 36).

During the Maunder Minimum, named after the English Victorian astronomer Edward Maunder, the eleven-year cycle of sunspot activity disappeared and sunspots almost ceased to appear on the disk of the Sun. During this lapse the Earth also suffered from what is often termed the "Little Ice Age," when the Thames River in England froze over, sometimes for days or weeks, and great ice fairs were held on its solid surface. Other anecdotes record how the "milk came frozen home in pail."

Chinese observations show similar, previous epochs when few, if any, sunspots were recorded with the naked eye for various decades at a time. These gaps suggest that the Maunder Minimum was not unique and was a natural part of the sunspot cycle, although we still have no idea how and why it happened. Nor do we understand why it should have ended.

Chinese records contain a number of observations of meteor showers as well, which have proved of great value in tracking the evolution of some well-known showers over two millennia. Some of these Far Eastern observations are surprising (see table 9.1). The first of the major meteor showers to be detected by the Chinese was the Eta Aquarids in 74 B.C., which, together with the Orionids, is the consequence of the repeated passes of Comet Halley

TABLE 9.1 First Date of Observation in Chinese Chronicles of Five of the Best Known Meteor Showers Observable Today

Meteor Shower	*First Observation*
Eta Aquarids	74 B.C.
Lyrids	About 500 B.C.
Perseids	July 17, A.D. 36
Orionids	A.D. 288
Leonids	A.D. 902

NOTES: Some other major showers seen in modern times, such as the Quadrantids, or the Geminids, have only existed for a few centuries at most (previously they did not cross the Earth's orbit) and could not have been observed by the Chinese. On the other hand, we also know that the Chinese observed some meteor showers that are unknown today but were highly active in the past.

that have been seen over the millennia. Each Orionid and Eta Aquarid meteor is a grain of dust expelled from the cometary nucleus which, probably thousands of years ago, formed part of a long-disappeared tail of this famous comet. Although the Eta Aquarid shower is now usually the much stronger of the two, neither of them can compare with the Perseids or Leonids at their best. It seems, however, that, in the past, some particularly spectacular showers of meteors may have occurred when the Earth met big concentrations of material from Halley's Comet, particularly in the case of the possible observation of an important Orionid meteor shower in A.D. 288.

Similarly, the Lyrids are normally a rather weak meteor

shower, with occasional stronger outbursts. In complete contrast to the activity (or rather, lack of it) that is seen most years when even an expert observer will rarely see more than ten or a dozen meteors per hour and the average person will not even notice that one is active, the Chinese observations, made on March 15, 15 B.C., note that "stars fell like rain." We know that, in recent times, there has been a much bigger Lyrid meteor shower approximately every 60 years over the last two centuries or so. The Chinese observations show us that the bigger showers also happened in the past.

Although the Far Eastern chronicles have been used to study many types of phenomena, where these Chinese records are supreme is in their records of comets and supernovas. While the Romans were discussing whether the Great Comet of A.D. 79 was an augury of the death of Emperor Vespasian or of the king of the Persians, the Chinese were recording its position, visibility, length of tail, and movement among the stars. Vespasian, quite naturally, preferred the second explanation of the comet's message, stating that "the comet refers to the king of the Persians because he is hairy and I am bald." In saying this, he was making play of the fact that a comet was often called a "hairy star." Chinese records show that the comet appeared in April of A.D. 79; it was seen first in the east and then moved to the north, remaining visible for twenty days.

In many cases, the Chinese observations are good enough to calculate at least an approximate orbit for a comet. In the case of periodic objects, such as Comet Halley and Comet Tempel-Tuttle, the observations enable us to calculate the comet's movement back in time over many centuries. This is simple to do. For example, we know the orbit of Comet Halley particularly well, but we also know that it changes slightly with time. Jupiter and Saturn modify the orbit—by speeding up or slowing down the comet—and the small jets of gas and dust expelled by the comet can accelerate or decelerate it as well. We can adjust the known orbit for these small changes until the period of visibility and position of the comet in the sky agree with the Chinese observations. Although these Chinese observations do not have a high degree of precision compared with modern data, at least as far as the position of the comet is concerned, they are more than sufficient to determine with great precision the moment of perihelion at each approach to the Sun.

By A.D. 635 Chinese astronomers were certainly aware that the tail of a comet pointed away from the Sun—a detail not "discovered" by Western astronomers until the sixteenth century. There is also some evidence that they realized a comet that disappeared close to the Sun in the morning sky would be the same comet that appeared, shortly afterwards, in the evening sky (and vice versa). Although they did not know the reason for this (that the comet had gone around the Sun and, after disappearing on one side shortly before perihelion, it would reappear on the other as soon as it had got far enough away from perihelion to be observable again), they knew it nonetheless.

The Chinese observations of Comet Halley in A.D. 141 are a particularly interesting example of the excellence of the Chinese as observers. Donald Yeomans, an astronomer for NASA who works at the world-famous Jet Propulsion Laboratory in Pasadena, California, and is the author of a widely respected book on comets, interprets the Chinese text as follows: "On March 27th (141 A.D.) a broom star was seen in the east with a tail about 9 degrees long and pale blue in color. The comet was seen until late April." Given this description, we know that the comet appeared for the first time just after perihelion, which had occurred on March 22.4 (i.e., around 9–10 A.M. on March 22). The pale blue color of the comet's tail is a rarely observed detail in naked-eye observations and refers to the presence of ionized carbon monoxide within the tail, which emits light of a strongly blue color. An estimate of the brightness of Comet Halley in A.D. 141, on the basis of its known distance and normal behavior, suggests that it was around magnitude −1 at the time of the Chinese observations. Calculations show that Halley's Comet passed particularly close to the Earth in that year. Sometimes, as in this case, the Chinese records lead to a small mystery because there is some detail that does not seem to fit. In this case, the calculations of the light curve of the comet show that it should still have been very bright indeed when last observed by the Chinese in late April A.D. 141. In principle, it would have been about magnitude 1.5 and well positioned in the evening sky for further observation.

Why did the Chinese not follow the comet for a longer period of time? Was Comet Halley actually a lot fainter than it should have been, on the basis of our estimate? Or was there simply a prolonged spell of bad weather in China that hid the comet from view? We do not know, though the former is more probable than the latter.

To back this up, we know that, on other occasions, the

Chinese observed Halley's Comet when we would have thought it was far too faint to be seen. Obviously it must have been a lot brighter than we expected. One example of such a case was the 1066 apparition, when the comet appears to have been particularly bright. However, the most spectacular case we know of was the return of Halley's Comet in 1145, when Chinese astronomers followed it with the naked eye until July 6, when it was well beyond the orbit of Mars and some 190 million miles from the Earth. This is the greatest distance from the Sun that the comet would be seen before the invention of the telescope. However, on July 6 the comet *should* have been well below naked-eye visibility, with its brightness around magnitude 6.5 (the naked eye can see objects that are about magnitude 6).

Such observations show us that the brightness of Halley's Comet changes from orbit to orbit. On a number of occasions in history, the Chinese observations have indicated that Comet Halley was either unusually bright or unexpectedly faint, which shows, almost beyond any doubt, that the comet's brightness must be variable from orbit to orbit.

A measure of the Chinese superiority as observers comes from the following breakdown of known observations of Comet Halley, from the first one ever made to the famous apparition in 1066 recorded on the Bayeaux Tapestry. In total, the comet was seen on seventeen occasions from 240 B.C. through A.D. 989 (see table 9.2), with two further possible, or probable, earlier sightings made by Chinese astronomers. As we can see, the Chinese failed to register Halley's Comet only once, in 164 B.C., which was the comet's worst-known appearance until 1986. Up

TABLE 9.2 Historical Observations of Comet Halley to the End of the Tenth Century A.D.

Total apparitions	17
Observed only in China	12
Observed only in Babylonia	1
Observed in China and Japan	3[a]
Observed in China, Japan, and Korea	1[b]

NOTES: Does not include the disputed observations in the fifth and twelfth centuries B.C. We can see how Chinese records dominate totally all other observations of the comet. Only on one of the seventeen occasions was the comet not observed in China; in more than 60 percent of cases, the only known observations were made in China.

[a]The first recorded Japanese observation of the comet occurred in A.D. 684. Since then, Japanese observations of comets have been comparatively common.

[b]The only recorded Korean observation of the comet before A.D. 1000 was in A.D. 989, which is, curiously, one of the poorest known apparitions of Halley's comet.

to the seventh century A.D., the Chinese observations are almost the only ones available. As we will see later, the fact that the Chinese were so superior to the Koreans and Japanese in observing comets may turn out to be an important clue about the nature of the Star of Bethlehem.

In table 9.3 we see the extent of a second and more unfortunate problem: the number of Chinese observations drops suddenly before the second century B.C. This is due to a cultural revolution that took place more than two millennia before the more recent one ordered by Chairman Mao. In 213 B.C. Emperor Qin Shih Huang ordered a "burning of the books," which was followed, just seven years later, by the sack of Xianyang, the Chinese capital at the time.

TABLE 9.3 Brief Summary of Chinese and Babylonian Records of Comets from the First Known Records in the Eleventh Century B.C. to the Time of the Fall of the Roman Empire

	Number of Comets Recorded	
Century	*In China*	*In Babylonia*
11th B.C.	2	0
10th	1	0
9th	0	0
8th	0	0
7th	2	0
6th	3	0
5th	6	0
4th	3	0
3d	6	2
2d	22	6
1st	15	0
1st A.D.	14	0
2d	19	0
3d	41	0
4th	17	0

NOTES: The increase in the number of comets observed by the Chinese in the second century B.C. probably reflects the greater degree of permanence and preservation of the Chinese records from this time, especially as many were lost in the "burning of the books" by the Emperor Qin in the third century B.C. The number of comets recorded by the Chinese is consistent (apart from the extraordinary number of comets registered in the third century A.D. with the current rate of appearance of bright comets of about one every three years. The relative paucity and severe restriction in time of Babylonian comets, all of which are found in the archives from just two centuries, suggests that many of their records are either lost or still untranslated; these missing records represent an enormous treasure house of potential information.

Despite the book-burning episode, the Chinese records of comets and probable comets include no fewer than sixty objects during the twelve centuries before Christ. In contrast, Babylonian astronomers register just eight comets in the same interval, although that number is still well ahead of those made in any other country. It shows that the Babylonians, too, were probably energetic observers, if less concerned with leaving permanent records.

The destruction of the oldest records in the archive means that many valuable documents from the Zhou dynasty, after 1050 B.C., a time when Chinese astronomy was particularly advanced, have been lost forever. But the more recent records that relate to observations made around the time of the birth of Jesus and may give us some clue as to the nature of the Star of Bethlehem are, fortunately, safe.

Chinese astronomers classified observations as being of various types. The three most common and best known are as follows:

1. *Hui-hsing.* Literally, "broom stars." These were bright comets with tails, which "swept" through the sky with their movement.
2. *Po-hsing.* Literally, "bushy stars." These were usually comets without tails, but they could also be a description of bright stars. Because of weaknesses in the human eye, bright stars seem to emit rays in all directions.
3. *K'o-hsing.* Translated as "guest stars," which occurred where no star had been seen previously. Usually these were novas or supernovas, but occasionally they could be misidentified (or mis-

> recorded) comets, or even meteors. In a few cases, *k'o-hsing* were variable stars seen by the Chinese at their maximum brightness and recorded as new stars. As the name implies, the *k'o-hsing* were generally regarded as being more auspicious than comets.

In 1977, David Clark and Richard Stephenson published a study of all the *k'o-hsing* that were observed by Chinese astronomers over the centuries. They found seventy-five records, some of which are easily identifiable as known events. The birth of the Crab Nebula in 1054 is faithfully reported, as is Tycho's supernova of 1572, although the latter was listed as a *po-hsing* rather than a guest star. A number of the Chinese records of guest stars are known to be almost certainly historical supernovas, but most of the rest were neither bright enough nor of long enough duration to have been supernovas and were, in the main, observations of normal novas.

A normal nova is a binary system of stars, with a white dwarf orbiting a larger star. This larger star may be a giant, or even a normal star. Material from the larger star falls onto the surface of the white dwarf and undergoes a "slow burn"; nuclear reactions occur, but only slowly, and are damped down by a layer of unburned material that has fallen on top of the layer that is burning. When the weight of the unburned material is insufficient to hold down the slow nuclear explosion underneath, the star suffers a crisis, and a quantity of material is blown off into space, clearing the surface of the white dwarf. The system is then stable once again, although sometimes a sufficient amount of new material accumulates to provoke a new crisis. There is a class of novas, called "recurrent novas," that suffer comparatively small explosions every few tens of years. For most novas, however, the interval is probably hundreds or even thousands of years between explosions.

A nova is quite unlike a supernova. The force of the explosion is much smaller, with a maximum luminosity of perhaps 50,000 times that of the Sun, compared to the supernova's 100 billion times. The blast wave from the white dwarf star is also not so strong; the wave expands outward at only 1000 miles per second, or a tenth of the velocity of a supernova blast wave. I say "only" because this velocity is still so high that if a nova were to go off in London, the blast would be felt in New York just a little more than three seconds later.

Inevitably, the Chinese chronicles and the observations from other countries in the region, usually referred to collectively as the Far Eastern records, have been closely scrutinized to see if there is any record of a comet, bright nova, or any other remarkable object close to the date of the birth of Jesus. These dates were not particularly rich in Chinese observations. Apart from the visit of Halley's Comet in 12 B.C. and a "ghost event" (i.e., a false record) in 10 B.C., only two events are found in the Far Eastern chronicles that are dated around the correct time. One of these is a Chinese observation, the other comes from a Korean chronicle. Both are extremely interesting indeed.

In a book called the *Ch'ien-han-shu,* we find the following reference: "In the second year of the period of Ch'ien-p'ing, the second month, a *hui-hsing* appeared in Ch'ien-niu for more than 70 days." If we convert the Chinese date to our modern calendar, the second year of the reign of Ch'ien-p'ing makes 5 B.C. the year when this event occurred. The second month in the Chinese calendar is equivalent to the period from March 10 to April 7. A *hui-hsing* is a broom star and Ch'ien-niu is the Chinese constellation that included Alpha and Beta Capricorni. Thus the complete translation would read: "During the interval between March 10 and April 7 of 5 B.C., a comet appeared close to Alpha and Beta Capricorni and was visible for more than 70 days." Curiously, the chronicle appears to state that the object remained fixed in the same place in the sky for more than two months. This would be very unusual if it were a comet.

In the Korean chronicles, a second record is found from about this date in the *History of Three Kingdoms—the Chronicle of Silla (Samguk Sagi)*: "Year 54 of Hyokkose Wang, second month, (day) Chi-yu, a *po-hsing* appeared in Ho-Ku." "Ho-Ku" is the Chinese constellation, or asterism, that includes Altair and various stars of the south of Aquila (the Eagle). A *po-hsing* is a bushy star: an extremely bright star with rays, or a tail-less comet. The problem here is the given date: the day Chi-yu did not exist in the second month of the year. It is almost as if the chronicler had written that the star had appeared on February 30.

One way to resolve this problem is to suppose that "Chi-yu" really should be "I-yu," a character written in an almost identical fashion and easily confused with it (fig. 9.1). If so, the chronicle really states: "Year 54 of Hyokkose Wang, second month, (day) I-yu, a *po-hsing* appeared in Ho-Ku."

Day "I-yu" of the second month of year 54 of Hyokkose Wang corresponds to March 31, 4 B.C. Thus the translation would be: "On March 31 of 4 B.C. a bushy star appeared close to Altair."

Figure 9.1. Calligraphic representation of the Chinese pictograms *Chi-yu* (*top*) and *I-yu* (*bottom*).

One of the greatest experts of recent times on variable stars is the Russian astronomer B. V. Kukarin of Moscow's Sternberg Institute. Kukarin led the team that compiled a catalog of 21,000 variable stars in the late 1960s and early '70s. This catalog, and its supplements, is still a standard reference work. Kukarin and his colleagues suggest that the 5 B.C. star in the Chinese account appeared within two weeks of March 24, 5 B.C., and was visible for at least seventy-six days. They point out that the monsoon season, which starts in April in the Far East, would have severely limited the Chinese capability to observe the object, hence seventy-six days is very much a minimum duration for the star's visibility. Their conclusion, however, which appears to be unique among researchers who have investigated this event: they think it was one of just four records prior to 1000 A.D. that appear to have been observations of Venus. This conclusion seems to be erroneous because Venus would have passed through the constellation of Capricornus at the end of 5 B.C. and would have been closest to the position of the Chinese star in November 5 B.C. This difference of six months appears to reject the Venus hypothesis.

For the 4 B.C. object the Russians give a much more specific translation, stating that it appeared on February 23, although there is no indication of its duration. They accept that this object was a nova.

To thicken the plot, Kukarin and his colleagues state that the 4 B.C. *po-hsing* was also observed from Palestine, but they give no reference to such an observation, nor do they explain the comment anywhere in their text. This looks like a veiled reference to the biblical reference to the Star of Bethlehem and suggests they feel that the 4 B.C. object *was* the Star of Bethlehem. If so, one can understand their reluctance to be more specific: their catalog was published in Moscow in 1971, under the Soviet system, which certainly would not have welcomed a reference to such a subversive text as the Bible.

In 1977 a group of British researchers—David Clark, John Parkinson, and Richard Stephenson—pointed out that this curious discrepancy in the date, due to the confusion between "Chi-yu" and "I-yu," leads to a strange series of coincidences:

1. Apart from the difference of one year, the date of appearance (March) is the same for both objects.
2. Both objects appeared in the same region of the sky; the asterisms of Ho-Ku and Ch'ien-niu border on each other, with Altair being located in the sky just a few degrees north of Alpha and Beta Capricorni (see table 9.4).

3. Both objects are described as cometlike (*hui-hsing* and *po-hsing*, respectively), but there is no indication that either of them moved across the sky in cometary fashion.

They concluded there were only two possible explanations for the two records:

1. They were really the same record, with the year also in error in the Korean chronicle. In truth, the object must have appeared close to the border between the two Chinese asterisms, which means it was located somewhere between Altair in the north and Alpha and Beta Capricorni in the south. The star would thus have been positioned, they concluded, quite close to the third-magnitude star Theta Aquilae.
2. A comet and a nova, or two bright novas, appeared in the same region of sky, around the same date, in consecutive years.

Of the two explanations, the first is much more plausible than the second, especially as the records that are being translated are transcriptions of the originals made several

TABLE 9.4 The Stars That Form the Asterism of Ch'ien-niu and Ho-Ku in the Chinese Star Maps

Asterism	*Principle Stars That Formed It*
Ch'ien-niu	Alpha, Beta, Xi^1, Xi^2, Pi, Rho, and Omicron Capricorni
Ho-Ku	Altair, Iota Aquilae

centuries later. The idea that two bright comets or novas could appear in the same part of the sky, on the same date, in two consecutive years, seems incredibly implausible. It also seems unlikely, given the known record of the Chinese as observers, that the Koreans but not the Chinese would record the 4 B.C. object, particularly as this was an early era for Korean observations.

Could it be that the two objects really were different? This seems rather unlikely at first glance, as it requires an amazing coincidence. Of the sixty comets listed in the book by Donald Yeomans, the NASA astronomer, as having been recorded by the Chinese in the centuries before Christ, only three were also recorded by the Koreans. What's more, there was no object that was seen by the Koreans but not by the Chinese. How likely is it that the Koreans would register an object and the Chinese not? Not very! Unfortunately, Yeomans throws a small wrinkle into this comfortable hypothesis by suggesting that the 4 B.C. object was indeed observed by the Chinese. He bases this assertion on the early catalog of the Far Eastern observations of Ho Peng Yoke, compiled in 1962. This is in direct contradiction to the result expressed by David Clark and his collaborators, which has since been strongly reiterated by Richard Stephenson.

The other highly debated point is the nature of the 5 B.C. and 4 B.C. "stars." Yeomans accepts both as comets, but Stephenson supports the suggestion that both refer to a single bright nova. In 1977 Clark, Parkinson, and Stephenson came down heavily in favor of the two reports being of the same nova and argue that the alternative suggestion of a comet is rather unlikely at least for the 5 B.C. event. David Hughes steers a middle course by referring to the 5 B.C. event as a comet and that of 4 B.C. as a nova!

To understand how different people can obtain such different conclusions from the same report, we must examine how vaguely some phenomena are described, even in otherwise detailed chronicles. For example, here we can draw some conclusions from the observations of Halley's Comet made in 12 B.C., which were particularly complete. To put the 12 B.C. observations in context, they refer to what was not a particularly spectacular apparition of Halley's Comet, although it passed just 14 million miles from our planet, inside and above the Earth's orbit, on September 10, 12 B.C., exactly one month before perihelion. The comet was first sighted on August 26, when it was just outside the Earth's orbit and probably about magnitude 4.5. The Chinese observations refer to it at that time as a *po-hsing*. During the fifty-six days that Comet Halley was observed, ending nine days after perihelion, the comet crossed the sky; the Chinese records track its journey across the heavens, from Canis Minor through to its disappearance in the glow of twilight when it was positioned in Scorpio, close to the brilliant star Antares. Despite its close approach to the Earth, the comet was never particularly bright and barely passed magnitude +1. In many respects, the circumstances of the comet and its appearance would have been similar to Comet Hyakutake in March and April 1996, although it would have appeared a little less bright than Hyakutake and would have probably shown a less well developed tail.

The seeming interchangeability between the aspect of comets and of bright stars shows up in many records, not just the Chinese chronicles. In his reference to Tycho's supernova of 1572, the account written at the time by Jerome Muñoz, a mathematician and professor of Hebrew at the University of Valencia (Spain), says that "I am certain that on the second day of November 1572 there was not this comet in the sky."

Even though he called the phenomenon a comet, later on in his account, Muñoz also refers to the object as a star. Chinese records also express this confusion with regards to Tycho's Star. The Ming-shih-lu refers to the 1572 object in the following terms: "It was as large as a lamp and pointed rays of light came out in all directions." The same account also calls the object a *po-hsing* or "rayed star," while the Ming-shih-kao calls it a *hui-hsing*, which normally means a comet but in this case clearly not. In other words, it was perfectly possible for the Chinese chroniclers to call an object a comet when it was really a bright star.

Clark and Stephenson have investigated in great detail the observations of all the objects referred to above, recorded by the Chinese, which might possibly have been novas or supernovas. They tried to use objective criteria to distinguish between comets, novas, and supernovas in the most reliable way and finally assigned a confidence factor to every object in their list of seventy-five candidates. Each of the seventy-five was rated from 1 to 5, with an object of class 1 being a long-duration stellar object, one that was an almost certain nova or supernova.

Of their candidates, just two are classed as *hui-hsing* and five as *po-hsing*. One of the former was the 5 B.C. object, which receives special attention from Clark and Stephenson. Because of its very long duration and the lack of any suggestion of movement, it is listed as a class 2 object: in other words, they regarded it as a probable nova or supernova. Of the twenty objects that were seen for at least fifty days, seven are probable or certain supernovas and an eighth is a serious candidate. Two more are certain or almost certain novas; five, including the 5 B.C. object, are in the "doubtful" category that includes possible comets. The 5 B.C. was included in this category not because of any allusion to motion (the usual reason for putting something in this category) but because of its description as a tailed object (*hui-hsing*).

The interpretation of the 5 B.C. *hui-hsing* as a comet leads to some interesting problems. The period of visibility is given as "more than seventy days." This is a particularly long period of naked-eye visibility, considerably greater than that of Comet Halley in 12 B.C. It suggests that the comet, if it was one, might have been particularly bright. Its position on the Aquila-Capricornus boundary implies it was approaching (or conceivably receding from) perihelion from a direction rather close to the Sun and, in consequence, from the Earth. Observers would have looked at the comet almost head on, so the tail of the comet would have been largely hidden behind. This would be consistent with the description of the object as a *po-hsing*, a bushy star, but less likely to cause a spectacular tail consistent with a *hui-hsing*.

This approach, from close to the Sun, also explains why the comet would not have moved much across the sky, for it would have appeared to come virtually straight toward the Earth. This direction also has other, important implications. First, the comet's distance from the Earth would have been very large and so, to appear bright, it must have been an exceptionally large object. Second, like Comet Hyakutake, even though it would have stayed almost fixed in the sky while it approached the Earth, it would have swiftly crossed the sky—hanging from being a morning object to an evening object—as it passed by. This was precisely the moment when the comet would have been brightest and the most spectacular. If it did not pass close to the Earth in this way, it is inconceivable that such a large comet would not be observed both before and after perihelion. What's more, the sudden change from being a morning to an evening "star" would almost certainly have been recorded by the Chinese.

Things don't change much even if we assume that the comet's orbit was arranged in such a way as to hide it behind the Sun for half of its orbit (in other words, it was seen only before or only after perihelion): during the more than seventy days of observation, the movement of the Earth would have caused the comet's apparent position in the sky to change substantially and would not have left it in a fixed or virtually fixed position. Only comets that are far from the Sun, or those heading virtually straight for the Earth can remain fixed for any period of time. The Korean and Chinese records, if they refer to the same object, leave little room for doubt; the object could not have moved significantly because the positions they both give are almost identical. In fact, they fix the position of the object to within a circle of about 5 degrees diameter, just ten times the apparent diameter of the moon.[2]

Common sense advises us that the simplest and best explanation is that both the Korean *po-hsing* and the Chinese *hui-hsing* refer to the same object observed in mid-March 5 B.C. and that it was most likely a bright nova and not a comet, although the comet option cannot be ruled out. The interpretation of a *hui-hsing* as a nova implies that the object would also be extremely bright, certainly of negative magnitude or close to it.

There is a counterargument to the nova theory, one that is important. Novas occur almost exclusively in the Milky Way, close to the plane of our galaxy. Therefore, the argument goes, if the object occurred at a latitude significantly above or below the plane of the galaxy, it is unlikely to have been a nova. Capricornus is at a rather high galactic latitude, the argument continues, which means it is an unlikely place to see a nova. This argument has weight; if true, it would tilt the balance strongly in favor of the 5 B.C. object being a comet.

It is true that Alpha and Beta Capricorni are some distance away from the plane of the galaxy (they are around 25 degrees south). Altair, however, is quite close to the plane, just 9 degrees south of the galactic equator and well within the visible band of the Milky Way. So, if we assume the 5 B.C. object to have a position somewhere between that given by the Chinese and by the Koreans, we get an estimated galactic latitude for this hypothetical nova of around 18 degrees south. If it were a bright nova, it is likely to have been comparatively close to the Sun in the galaxy; this means it could *appear* to be some way away from the galactic equator even though it was actually close to the galactic plane. This is backed up by the observed positions of the naked-eye novas and supernovas seen in the last two millennia. A total of sixty have been observed, distributed as shown in table 9.5.

Among the novas observed at galactic latitudes greater than 15 degrees, we find two first-magnitude and two

TABLE 9.5 Distribution of Galactic Latitude for the Sixty Naked-Eye Novas and Supernovas Observed to 1986

Latitude Range	*Number Observed*	*Percentage*
≤ 5 degrees	28	47
6–15 degrees	24	40
16–25 degrees	3	5
26–35 degrees	3	5
36–45 degrees	0	0
46–55 degrees	2	3
≥ 56 degrees	0	0

NOTE: One in eight of all novas appear at more than 15 degrees above or below the galactic equator. In other words, a bright object close to the plane of the Milky Way is most likely to have been a nova, although a significant number do occur some distance away from the Milky Way.

second-magnitude objects. These include such important objects as Nova Herculis of 1934 and Nova Puppis of 1926.[3] It is by no means impossible, therefore, for the object to have been a bright nova—provided it appeared between the two asterisms quoted in the Far Eastern records. In fact, of the sixty bright novas and supernovas listed from the Christian era, no fewer than 6 (10 percent) have appeared in the constellation of Aquila, although no bright nova has ever been seen in Capricornus.

It is also possible that the 5 B.C. object was a "fast nova." This name is given to the class of novas that decline rapidly in brightness. Just occasionally a bright nova may, in extreme cases, go from being a bright naked-eye object to being completely invisible to the naked eye in a week or less. The Chinese star may have been a fast nova, because ordinary slow novas tend to have erratic light curves, often with several separate maxima. A fast nova may fall five or six magnitudes in seventy days, while a slow one can be close to maximum for a full year. In fact, a slow nova may hardly fade at all in seventy days and, in extreme cases, might not even reach its maximum brightness during that time.

We are led to conclude that a bright object, probably a nova, appeared in mid-March 5 B.C., probably quite close to the star Theta Aquilae. The position of the nova would have been around Right Ascension 20h00m, Declination −03 degrees. The nova was visible for around two and a half months and, although we cannot really know how bright it was, we can guess that it must have been at least magnitude zero, and possibly much brighter. This nova would have blazed in the dawn sky, quite low in the east.

The date, March 5 B.C., and the position of the object in the sky, quite low in the east at dawn, are highly significant and fit in with the date of appearance and position in the sky of the Star of Bethlehem. The agreement is close enough to suggest that this object, in some way, shape, or form, was indeed the Star of Bethlehem. Perhaps the answer to the mystery of the Star was in fact written in Chinese, as the Soviet scientists hinted in 1971.

10

What Was the Star of Bethlehem?

Over the nearly two millennia that the debate about the nature of the Star of Bethlehem has lasted, dozens of theories and explanations have been proposed. Some experts do not believe the Star ever existed, compelling those who seek to explain it to show convincingly that it *did* exist. Although the situation is not quite as bad as in the case of the death of the dinosaurs—almost one hundred different theories have been published to explain their extinction—there are still many theories to consider. One would think that, as with the dinosaurs, much of the debate is based on sound (if not conclusive) physical evidence. This, however, is just not true. We have seen that there is little or no physical evidence; those who study the death of the dinosaurs have, if nothing else, a glut of fossils to examine. Those who study the Star of Bethlehem have few "fossil records" and must be resourceful and able to make deductions from very limited supplies of evidence.

This is the reason why the mystery of the Star of Bethlehem has not yet been solved. Perhaps it never will be, nor can it be resolved in a strict sense of the word, if we take "resolved" to mean proved beyond all reasonable doubt. We can, however, make a lot of educated guesses and we now know many things that were not known even a few years ago.

Modern science has resources that could not have been imagined a century ago. New theories have been proposed and old ones modified. Evidence is found in the most unexpected places. Clues to the nature of the Star of Bethlehem may be found, as we have seen, in Italian churches and chapels and in the sands of the desert, in Chinese pictograms written on animal bones and on silk, and in a gold mine in South Dakota. A scribe's spelling mistake and an apostle's error have all combined to veil the truth. In many places, there are false trails that lead to dead ends.

To explain the Star of Bethlehem, if it really did exist, we have to satisfy a number of criteria:

1. The explanation must be compatible with the probable date of the Nativity.
2. It must be a singular, special, or spectacular event.
3. It must be a rare event.
4. It must have had a special meaning for the Magi.
5. It must have occurred in the east.
6. It must have endured for some time.

Many of the explanations suggested in the past are unconvincing, or unsatisfactory because they fail one or more of these criteria. It is unsatisfactory to try to explain the Star of Bethlehem using an event that is particularly common or repeats itself with too great a frequency: something that occurs just once a century may sound as if it is a rare event, but it becomes a common one if we are talking about a timescale of a thousand years during which the birth of Jesus was anticipated. Given that the Magi and their ancestors had watched the sky for at least a few centuries and possibly for a millennium, a once-per-century event that seems special in a narrow perspective suddenly becomes much less special. Such an event could have sent the Magi (or their great-great-grandparents, or their great-great-grandsons) to Bethlehem a hundred years too soon or a hundred years too late. So we are searching for something that was totally unequivocal and would not have been expected to be repeated in any reasonable time frame, or the Magi would have stayed home.

We must tie in the rarity of an event with the special significance that any phenomenon invoked to explain the Star of Bethlehem must have. This does not imply that a "common" object could not have been the Star of Bethlehem. A common event, such as an occultation or a comet, could be regarded by the Magi as *the* sign, provided that it occurred in a particular way. Given that the Magi were astrologers first and foremost, they would look at any potential sign in a special way. It is possible that we could never fathom their thought processes, for a thing that appears insignificant to us now may at the time have been of overwhelming importance to them for reasons we cannot now imagine.

This argument has been made in various ways. A conjunction in the constellation of Leo would be significant because Leo was a royal constellation, and a conjunction there would indicate a royal event. A conjunction in Pisces would indicate an event in Judea because Pisces was the constellation associated with the Jews. A nova or a supernova would be significant because a new star would be a portent of a birth, and a bright new star could only mean the birth of a king. A comet would be a significant portent for reasons similar to the nova, although it is also true, as some authors state, that a comet is an unlikely Star of Bethlehem because comets were (and still are in some parts of the world) associated with deaths rather than births. The occultation of a bright planet by the Moon could also be a portent because the planet would disappear (die) and then, a short time later, be reborn.

All of these celestial events were seen at some time within a few years of the date of the Nativity. If all were important enough, or significant enough to have been the Star of Bethlehem (and each one has its supporters), we can only suppose one of two things: either the Magi crossed the desert, back and forth, at least once a year until they got the right sign; or there was an additional circumstance that made one event really special. With so many portents in the sky, choosing just one must have been virtually insuperable.

We have an advantage over the Magi because, with the benefit of hindsight, we know that the date of the Nativity was probably around March–April 5 B.C. We can thus filter out all events that were not visible within a small interval of time around that date; the Magi could not do this. This is a particularly important point, one that the theories of the Star of Bethlehem rarely, if ever, address. It is one that we must not just address, but we must also satisfy it.

The Magi had been watching the sky patiently for hundreds of years and had seen thousands of conjunctions, dozens of comets and bright novas, maybe even a few brilliant supernovas. None of them had been sufficient to be the sign that they sought unless we assume the Magi had only recently started to observe the skies and that the first great sign they saw was the Star of Bethlehem itself. But this does not seem to fit even with what little we know about the Magi. So why, after all these years, should they get excited now? Any of the multiple explanations that researchers have presented fails to clear this final hurdle because no one can locate an astronomical event or a phenomenon recorded in ancient documents so spectacular *on its own* that it was clearly the Star.

A particular case in point is Halley's Comet. The comet is neither a particularly bright one nor a rare event. It was a plausible explanation for the Star of Bethlehem only because we now know that it appeared close to the date of the Nativity. With an average of four very bright comets appearing every century, if the Star was a comet, we must ask ourselves why the Magi did not visit Jerusalem once every twenty-five years, every time they saw a bright comet. Conjunctions also fail to satisfy this criterion. Even now, two millennia later, we cannot decide beyond all reasonable doubt which conjunction seen in the few years around the birth of Christ was the best candidate for the Star of Bethlehem. What's more, *we* know roughly when the Star must have appeared because we know the date of Jesus' birth, but how could the Magi have known what event was the *real* sign? With so many conjunctions to react to, they would have required an air shuttle service rather than camels to travel to possible birth sites, collecting frequent flier miles at a staggering rate. In the centuries that the Magi had watched the skies they would also have seen several brilliant supernovas and many novas. If the "Star" was a nova, why should the Magi single out a particular one as worthy of their attention? A meteor is no better as a candidate because we must rely on two particularly bright or unusual ones, each appearing at just the right times separated by a few weeks. Even if we search for an event with a particular astrological significance, within so many centuries any one event would eventually be repeated, making it not so unique.

If we are to explain the Star of Bethlehem as an "ordinary" astronomical event, what we need is an event, or a *combination* of events, so unusual, so remarkable that its significance would be obvious, at least to the Magi. As we have seen, no single event stands out. The fact that there are so many possible candidates shows that, almost certainly, none is sufficiently important to be the right one *on its own*. We can thus reject the suggestion that only a single object was involved. The only possible solution to the mystery, if we accept the notion that the Star did exist, is a combination of events. What series of events would have been of great importance and significance to the Magi but not necessarily to anyone else?

To answer this, let's look more closely at three interesting but in no way unusual astronomical events that occurred between 7 B.C. and 4 B.C.:

1. The triple conjunction of Jupiter and Saturn in Pisces between May and December of 7 B.C.
2. The planetary massing of Mars, Jupiter, and Saturn in Pisces in early 6 B.C.
3. The object or objects observed by the Chinese and Koreans in Aquila/Capricornus in the spring of 5 and/or 4 B.C. (which are assumed to be one and the same).

To these we should probably add the occultations of Jupiter by the Moon in 6 B.C. as a more recently discovered, but still highly important, astrological event.

The 5 B.C. object is a very interesting one. If we accept that the date of the Nativity was late March or early April of 5 B.C., the coincidence of timing between the Chinese *hui-hsing* and Korean *po-hsing* and the date of the Nativity is complete and absolute. If the Nativity really occurred in March 5 B.C., it would have coincided in date with the appearance of the Chinese star. It is hard to believe, in this case, that we need to look any further for the Star of Bethlehem. Just by the coincidence in dates, the 5 B.C. object was—*must* have been—the Star.

But this is far from the full story yet. This theory still suffers from the same fatal problem of all the others: Why should the Magi think that this object was any more significant than any other nova they had seen over the years? The only explanation is that they already knew when Jesus would be born and so could react at the correct time.

The Magi's attention had been drawn by a series of events that had left them in a state of anticipation. They knew that something was afoot but were waiting for a final, definitive sign. We can imagine that, over the years, time and again, the Magi had seen signs in the sky that led them to believe something was going to happen in the Holy Land. Time and time again, though, they were disappointed because the *final* augury had not appeared. This time, however, after many years—perhaps generations—of waiting, a whole sequence of events had happened and they saw everything they had waited for. This time they knew this was it. What the Magi had seen was a chain of four closely connected signs, each of which was highly significant to them.

The First Sign: A Triple Conjunction in Pisces

In 7 B.C. they saw the triple conjunction of Jupiter and Saturn. It was not a particularly spectacular one, admittedly, for the separation between Jupiter and Saturn was still almost a full degree at minimum. But it was a significant conjunction because of the way it occurred. Early in May they would have seen the two planets start to approach each other in the sky. On May 29 they passed each other, one slightly less than a degree to the north of the other in the constellation of Pisces. Knowing that Pisces was the constellation associated with the Jews, they would have followed this development with some interest, though probably they were not overly excited because they had seen something similar many times before. The fact that Jupiter, a royal planet, was involved might have caused some extra interest, but this is far from certain because there are many conjunctions of two or more

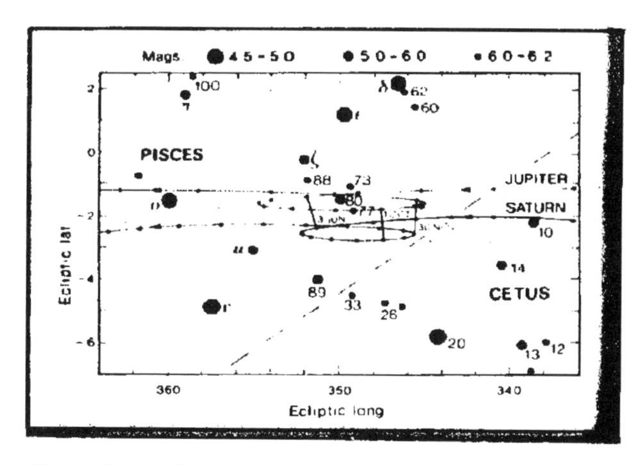

Figure 10.1. The triple conjunction. (Patrick Moore, from *Astronomy Now* magazine.)

planets that involve Jupiter. Through the late spring and early summer they would have seen how Jupiter and Saturn separated in the sky and, in the absence of anything else of interest, might have felt slightly disappointed that a real sign had not yet been given. This was soon to change (see fig. 10.1).

In early August they would have noted with increasing excitement that the two planets had stopped separating and had, slowly, very slowly, started to approach each other again. By early September there would have been no doubt: the two planets were approaching each other much faster now and a new conjunction was in the making. On September 29 the two planets again passed each other, one to the north and one to the south, moving in opposite directions. This time the separation between the two was a little larger, but the conjunction had occurred once again—in the constellation of Pisces. As the month of

October passed, the Magi would have continued to note how the two planets separated, although they would by now have been watching closely for further developments. This time they did not have long to wait: after just a single month, the separation of the two planets stopped for the second time. With growing excitement, the Magi would have seen how, for a third time, the two planets started to approach each other again! Through the month of November, Jupiter and Saturn would have come closer and closer together, although still separated by more than a degree in the north-south line. Finally, on December 4 the third conjunction occurred. This time the two planets separated slowly but steadily.

The Magi would have stopped to think about this. Three times in six months the royal planet and Saturn had met and then separated in the constellation of the Jews. Surely this meant that something important was about to happen in Judea. The fact that a royal planet was involved suggested that a royal event was imminent: A king would be born? One would die? King Herod was now an old man, hanging on to life—perhaps the sign referred to him? Saturn was regarded as an evil planet, and maybe this increased the Magi's suspicions that the detested puppet leader of Judea was somehow involved. The Magi would have pondered long and hard about the possibilities but, remembering previous disappointments, would have waited to see what else might happen.

The Second Sign: A Massing of Planets in Pisces

Once again, they would not have long to wait. As Jupiter and Saturn slipped down in the evening sky toward the horizon and, night after night, became harder to see in the bright twilight, another event was happening. Mars, which had only a few months earlier been far away in the sky, at opposition, started to enter the constellation of Pisces.

In February of 6 B.C. the three planets, Jupiter, Saturn, and Mars, were separated by only about 8 degrees. This event, a massing of planets, also happened in the constellation of Pisces. Even though it would not have been a particularly spectacular event, its astrological significance would have been great and its impact tremendous.

The Third Sign: Two Pairings in Pisces

But something even more interesting and eye-catching was about to happen. On the evening of February 20, 5 B.C., the two-day old Moon would have passed very close to Jupiter in the sky. The Magi would have seen the spectacular pairing of Jupiter and the Moon slightly to the east of yet another pairing, of Mars and Saturn. This configuration, in particular, would have intrigued them. The Magi would now have seen four events occurring in Pisces—the constellation they associated with the Jews—in quick succession. The Magi knew that something was going to happen in Judea.

Let's speculate. One possible interpretation of the events would be the following. Jupiter is a royal and benevolent planet, while Saturn is malign and Mars invokes thoughts of war. The encounter between Jupiter and Saturn could have suggested to the Magi that a great ruler (the awaited Messiah) would arise, challenging a malign one (the Roman Empire), and liberate his country by the sword (as signified by the bloody color of Mars). This interpretation would have been in tune with what they would have expected from the oracle of Balaam:

> I see him, but not now; I behold him, but not near—a star shall come forth out of Jacob and a sceptre shall rise out of Israel; it shall crush the borderlands of Moab and the territory of all the Shethites.

The Magi would have waited, patiently, hopefully, for the Star to tell them when this was going to happen. The stars, in this case in the form of the planets, had given them a message: "Look toward Judea and await further developments." After so many years of watching the skies and waiting, the Magi would have been resigned to waiting just a little longer to find the answer to their question.

Shortly after the planetary massing came the twin occultations of Jupiter by the Moon. These events were probably known to the Magi, even if they did not observe them. There does not appear to be a reference to them in the Babylonian tablets, but that may be a false trail. We know that the surviving Babylonian observations of the sky are fragmentary, and we have seen that those may not even be the most important ones. We do not know if the Babylonians or Persians could predict the motion of the planets to some degree. If they could, they might easily have been able to calculate that an occultation was due. Astrologically speaking, these occultations would have provided the strongest clue yet as to the nature of the events that were brewing in Judea. The occultation would have clearly foretold the birth of a king in Judea, and their knowledge of the Jewish scriptures and prophecies would have suggested strongly who the king would be.

All they needed was a clear sign, something that, unlike the occultation, would be easy to see, unmistakably written in the stars. For a full year they waited patiently, prepared, once more, for disappointment. On this occasion, though, the stars would answer their prayers, and the fourth, final, and definitive sign indicating the birth of the promised Messiah would indeed come.

The Final Sign: A Nova

In February or early March of 5 B.C., about a year after the massing of the planets and less than a year after the lunar occultation, a nova blazed over the border between the constellations of Capricornus and Aquila. This nova was seen in the eastern sky in the first light of dawn. The Chinese records suggest, from its duration of visibility, that it was quite bright, perhaps even very bright. The designations *po-hsing* and *hui-hsing* normally applied to comets were also, as we have seen, given to very bright novas and supernovas.

On first sighting the nova, the Magi would have known their wait was at an end. The conjunctions had told them to await news from Judea and, possibly, to expect the imminent birth of the Messiah. The occultations told them that the new king was indeed the Jewish Messiah. The nova now told them that the royal birth had finally happened. The final piece of the puzzle was in place, and the Magi would only have had to act.

The logical destination for their journey to find the newborn king was Jerusalem, the capital of the Jewish world. Reaching it would require a long traverse of an inhospitable desert; they would have to make careful preparations. But compared to other journeys made regularly by camel trains, the 540-mile trip to Jerusalem from Babylon, or even the much longer traverse from Persia, would not be a particularly long one.[1] This one, though, was to be the most important desert crossing of their lives. The Magi knew that the journey would take several weeks of travel, that they would rise before dawn and ride through the morning. They would stop and retire to their tents to sleep peacefully and comfortably through the scorching heat of the midday sun, rising again in the late afternoon, after the worst of the heat had passed. In this way they would best conserve their precious supplies of water. Even so, a week, two weeks at the most, would have sufficed to prepare for the journey and to depart. What are a few more days when one has waited so many years?

Apart from the rigors of the Syrian desert, most of the journey would be comparatively easy: leaving Babylonia in a generally westerly direction, their path would have climbed from near sea level slowly to a level of over 3,000 feet. There were, however, no mountain ranges to cross, no rivers to ford, and only the endless sands and the range of hills extending southward from Syria to slow them down. Even if they only traveled for six hours per day they would arrive comfortably in seven weeks. Assuming that the urgency of the situation required more speed, the Magi could have ridden far more hours per day; in this case, it is probable that four to five weeks would have sufficed. In a maximum of six weeks, perhaps five, from the moment of apparition of the nova, the Magi would have arrived in Jerusalem to seek their audience with King Herod.

Had the Magi come from Persia, as we ncw suspect, the journey would have been more complicated. As we have seen, the traverse from Persia is about twice as long and would include a crossing of the Zagros Mountains before reaching the fertile plain of the Tigris and Euphrates Rivers. The whole journey would have taken much longer, and it would have been more difficult for the Magi to arrive comfortably within the seventy-day period of visibility of the object seen by the Chinese and Koreans. But it is far from impossible.

We know that the star observed by the Chinese and Korean astronomers in March and April of 5 B.C. was visible for "more than seventy days." More than seventy days implies a minimum of ten weeks, and the Magi would have taken no more than six weeks to arrive at Bethlehem from Babylon. Provided that the Magi were able to set out within two or three weeks of the apparition of the nova, they would have had no problem following it during their entire journey. With a little more hurry, it is perfectly possible that the Magi would have reached Bethlehem even from Persia within our time limit. A journey lasting ten weeks implies a speed of only about 15 miles per day, well within the capabilities of a camel, as I know from personal experience.[2]

As visiting dignitaries, the process of obtaining the audience would have been reasonably rapid and straightforward for the Magi. Herod would not wish to offend important visitors from a foreign power and would certainly have offered them hospitality and lodging while they recovered from their journey and made preparations to continue southward. After a week of rest, the purchase of supplies, the business of diplomacy, and consultations with Herod, the scribes, and Pharisees, they would know where to find the newborn child they were seeking.

The most interesting aspect of this scenario is that the Star, which was in the east when first seen, would no longer be in the east upon their arrival. Every two weeks the Star would have risen an hour earlier until, two months later, it would have been almost exactly in the south at dawn. When the Magi set out for Bethlehem, they would have seen the Star before them in the south at dawn. It is not necessary to assume, as many have done, that the Star had to have been a comet simply because it moved from the east to the south—this is the natural motion of the heavens! The Magi would, in the most natural way of all, have been able to follow the Star in the journey south from Jerusalem to Bethlehem, provided that they traveled at dawn, which they probably did.

From Bethlehem, at latitude 31.3 degrees north, a nova in the position indicated by the Chinese and Koreans, around Declination −10 degrees (that is, 10 degrees south of the celestial equator), would have been around 50 degrees high in the south at dawn. The Star would not have been in the zenith, because it would have passed through the zenith only when observed from slightly south of the equator. But it would have been high enough in the southern sky to appear above Bethlehem as the Magi approached the town.

There has also been the suggestion that the Magi lost sight of the Star before setting out for Bethlehem, hence they rejoiced on seeing the Star again over the town. Once again, this has a simple and very logical astronomical explanation. The Moon passes once a month through the constellation of Capricornus, as it is a zodiacal constellation, and its light would hide the Star as it did so. Although the Moon would not occult the Star, according to its phase at the time of the conjunction with the Moon and according to the brightness of the nova, the Star might be hidden for several days, or even a week. The tremendous brightness of the sky around the Moon would make the Star invisible to the naked eye unless the Star itself were extremely bright. If the Star were a normal nova, it is most unlikely that it would still be very bright so many weeks after first appearing. By the time the Magi reached Bethlehem it should have been starting to fade quite significantly.

Besides the monthly close approach of the Moon to the nova, another factor is involved here. Initially, with the nova comparatively low in the dawn at sunrise, the encounter between the Moon and the nova could only happen when the Moon was a small, waning crescent. Two months later, though, this encounter would occur in the south with a much brighter last-quarter Moon. The Moon's glare would hide the Star for possibly several days at a time. Also, once a month, at Full Moon, all but the brightest stars are hidden from view. With the nova close to its maximum brightness, this would not have been a great problem: what is more, the Full Moon would be setting in the west as the nova was rising in the east. Later though, as the nova began to fade and rose earlier and earlier each night, the Full Moon would make its observation very difficult, or even impossible, even though the Star was not close to the Moon in the sky. In other words, it is to be expected that the Magi would lose sight of the Star for a few days, or even a week, each month:

either when close to the Moon in the sky, or close to Full Moon.

In March 5 B.C., the Moon would have been new on the first day of the month and, again, on the thirtieth. In between, it would have passed close to the nova only as a thin, waning crescent late in the month, which would have had little effect on the nova's visibility. In April, though, the situation would change quite dramatically. Between April 20 and 21 the last-quarter Moon would have been in Capricornus and quite close to the nova, which would have been virtually invisible unless it was still very bright. Even a week earlier, as the Full Moon flooded the sky with light, it is probable that the nova would become hard to see. In May, when the nova was lost, it would have been close to a Full Moon in the sky and, now fading greatly, invisible for several days, or for even a week around the thirteenth, due to the Moon's brilliance. Add some bad weather and a few nights when the stars would have been invisible because of clouds, especially just before or just after either Full Moon, and it is not hard to understand how the Star could have been lost from sight for a time that was disturbingly long for the Magi.

By six weeks after maximum light, the nova would have faded by three, four, or even five magnitudes (a factor between 15 and 100 in brightness). It may even have faded more depending on the type of nova it was. What had been a brilliant object in the dawn sky would now be a much fainter star, still visible in a dark sky, but none to easy to see with a bright Moon, or even in a thin veil of cloud.

Why did the Magi travel to the Holy Land on this occasion and not on previous ones? The reason is very simple: the nova, without the prior conjunctions, would have had no special significance; the conjunctions, without the nova would have had no special significance. Only when a bright nova closely follows a particularly significant conjunction are the circumstances of a *special* event in place. A "bright" nova (maximum around magnitude 2) occurs approximately every twenty-five years. Such a nova would be noticed, but would not normally draw much attention to itself.

In contrast, a very bright nova (maximum in magnitude zero or brighter), which could reasonably be called a spectacular object and which would be consistent with the rather long period of visibility given by the Chinese, is much rarer: it occurs just once a century at most. A triple conjunction occurs every 150 or 200 years. The probability of a bright nova and a triple conjunction occurring within sixteen months of each other is very small. And the probability of both occurring together just at the time of the Nativity is less than one in ten thousand. This probability is so small that it is almost nonexistent.

If we include the other observed factors, such as the planetary massing (not all triple conjunctions give rise to one) and the fact that the nova occurred fairly close in the sky to where the triple conjunction had already been seen, we find a series of events so unique that they can happen together only once in every several thousand years.

9 When they had heard the king, they set out; and there, ahead of them, went the star that they had seen at its rising, until it stopped over the place where the child was.

10 When they saw that the star had stopped, they were overwhelmed with joy.

11 On entering the house, they saw the child

> with Mary his mother; and they knelt down and paid him homage. Then, opening their treasure-chests they offered him gifts of gold, frankincense and myrrh.

For those who wish to find a purely scientific explanation for the Star of Bethlehem, something that makes the event truly unique, this combination of events, culminating with a bright star, is compelling. In this case, the skeptics who believe that Matthew added the account of the (nonexistent) Star for religious and not historical reasons are wrong and there *was* a Star of Bethlehem, which was indeed a star and not a comet or a conjunction of planets, even if we cannot now identify it with a telescope. As we will see, though, we may yet be able to identify the Star for certain.

On the other hand, for those who look to the Star of Bethlehem for deeper meaning, who can say to them that such a rare combination of events—several conjunctions and a bright nova blazing brightly in the east at the moment of Jesus' birth—was not indeed miraculous?

EPILOGUE

Which Star Is the *Star?*

Where is the star now—once a bright object, almost certainly a nova—that appeared in mid-March 5 B.C. somewhere between the modern constellations of Capricornus and Aquila? If it was a nova within the region of the sky consistent with Chinese records, it most likely would have appeared slightly farther north of the division between the two constellations—that is, closer to the center of the visible band of the Milky Way—than it would have been if it had appeared within Capricornus itself.

Some years ago I wondered if it might be possible to identify a possible historical nova close to this position by searching for a variable star of the right type nearby. Binary star systems that give rise to a nova usually show some kind of variability in light, even when there is no nova outburst.

There is still another possible, more clear-cut way to identify the star. As we have seen, although we cannot prove it is true in all cases, nova explosions repeat themselves over the years. There are even so-called recurrent novas, which show many small explosions and have been seen to erupt time and again. The outbursts of a recurrent

TABLE E.1 Four Examples of Recurrent Novas that Reach Naked-Eye or Near Naked-Eye Visibility

Star	*Years of Outburst*	*Brightest Magnitude*
T Coronae Borealis	1866, 1946	2.0
RS Ophiuchi	1901, 1933, 1958, 1967	5.1
T Pyxidis	1890, 1902, 1920, 1945, 1965	7.0
WZ Sagittae	1913, 1946, 1979	7.0

NOTES: Typically, the more frequent the outbursts are, the less the star will increase in brightness during a nova outburst. T Coronae increases around nine magnitudes (a factor of 4,000 increase in brightness), while the more frequent outbursts of T Pyxidis are of only seven magnitudes (a factor of 600 increase in brightness). A normal nova may increase in brightness by fifteen magnitudes (a factor of a million) or more.

nova may be separated by as little as a few years (see table E.1). Full-fledged novas may erupt at intervals of many hundreds or even thousands of years.

Unfortunately, there are many variable stars in the constellations of Aquila and Capricornus. The *General Catalogue of Variable Stars* lists all known variable stars in the sky. It includes, even in its 1969 edition, 1,182 variable stars in Aquila and 61 in Capricornus. The ratio of novas in the constellations is likely to be similar, and they are thus twenty times more likely to appear in Aquila than in Capricornus.

Given so many candidate variable stars in a small region of the sky and our rather poor knowledge of the position of the 5 B.C. star, it is definitely easier to limit the search to recent novas, just in case one of them happens to be a repeat outburst of the putative 5 B.C. nova explosion.

Several stars in Aquila, which are in approximately the

right position, are listed as "nova" or "nova-like." Any of these could have been the Star of Bethlehem. Unfortunately, none of them appears to be really close to the most likely position of the star observed by the Chinese. The closest is a star known by its catalog name of DO Aquilae, although even this one is a little too far to the northwest. DO Aquilae was discovered as a ninth-magnitude nova in 1925, but is normally a very faint star of magnitude 18, visible only with a moderately large telescope. Such a nova is termed a "slow nova," meaning that it took a particularly long time to decline from maximum. DO Aquilae is at galactic coordinates of longitude 031.7 degrees, latitude −11.8 degrees. Clarke and Stephenson, however, initially estimated the coordinates of the 5 B.C. star to be longitude 030 degrees, latitude −25 degrees. However, they later suggested that their position is too far south, certainly by 5 degrees, and possibly by as much as 10 degrees. The galactic longitude of DO Aquilae agrees quite closely with that of the most likely position of the 5 B.C. star, although the galactic latitude is somewhat farther north.

If we take into account the estimate of the possible error in the latitude of the Chinese star given by Clark and Stephenson, the true latitude may have been as low as −15 degrees. This means that the positions of DO Aquilae and that of the 5 B.C. star are not necessarily very different. In fact, in a best case, the agreement between the positions of the two is particularly close, as shown in table E.2.

If the nova did appear between the two asterisms referred to in the oriental texts, as Stephenson later suggested, its position would have been within just a few

TABLE E.2 The Most Likely Position of the Chinese Star Observed in 5 B.C. Compared with the Position of the Now Faint Star DO Aquilae that Gave Rise to a Nova Explosion in 1925

	Galactic Longitude	*Galactic Latitude*
5 B.C. star	30 degrees	−15 degrees (?)
DO Aquilae	31.7 degrees	−11.8 degrees

degrees of the star DO Aquilae. The difference in position, in the best possible case, could have been as little as 3.5 degrees. In other words, there is a chance—maybe even a moderately good chance—that the Star of Bethlehem could have been a previous eruption of this nova. The position of this star in the sky is one-third of the distance from 42 Aquilae to 26 Aquilae, two quite faint (almost sixth-magnitude) stars, which are a little to the south of the star Iota, one of the main stars of the Chinese asterism. Both 42 and 26 Aquilae are none too easy to see with the naked eye and difficult to locate without a star chart.

"Dawn," the time of day when the Magi's star was seen, is defined in various different ways by astronomers. "Civil twilight" is better known to motorists, at least in the British Isles, as "lighting up time"—the moment when car headlights should be used at dusk and can be switched off at dawn. This is when the Sun is just 6 degrees below the horizon and the whole sky is bright enough to hide all but the brightest stars. "Nautical twilight" is the name given to the moment when the Sun is 12 degrees below the horizon. At this time there is a broad band of light where the Sun has set, or where it will rise, but most of the sky is dark and filled with stars. Finally, there is "astronomical twilight," when the Sun is 18 degrees below the horizon. At this time the sky is completely dark apart from the first glimmer of light on the horizon in the morning and the very last glimmer at dusk.

So when we ask where the Magi's star was at dawn, it is an inexact question and can have several equally valid answers.

At nautical twilight on February 20, 5 B.C., the star DO Aquilae would have been 39 degrees above the horizon, in the southeast, having risen two and a half hours previously, at 2:17 A.M. local time (fig. E.1). Even at astronomical twilight, with the first faint glow of light on the horizon, the star would have been just over 30 degrees high, beautifully visible in the dawn sky.

One problem with this possible association between the two objects is that the observed outburst of DO Aquilae in 1925 was a rather small one; to be a plausible Star of Bethlehem, we would have to assume that the hypothetical previous outburst was perhaps five thousand times larger. But perhaps this is not as big of a problem as it might appear. If DO Aquilae really did have a huge eruption in 5 B.C., it is quite possible that this explosion removed the need for a further big eruption of the star for a very long time: successive ones might be smaller until, once more, the system reaches a new and massive crisis.

It is certainly a long shot to suggest that this faint star, DO Aquilae, was the one seen more than two thousand years ago by the Chinese astronomers and by the Magi—but it is not completely impossible.

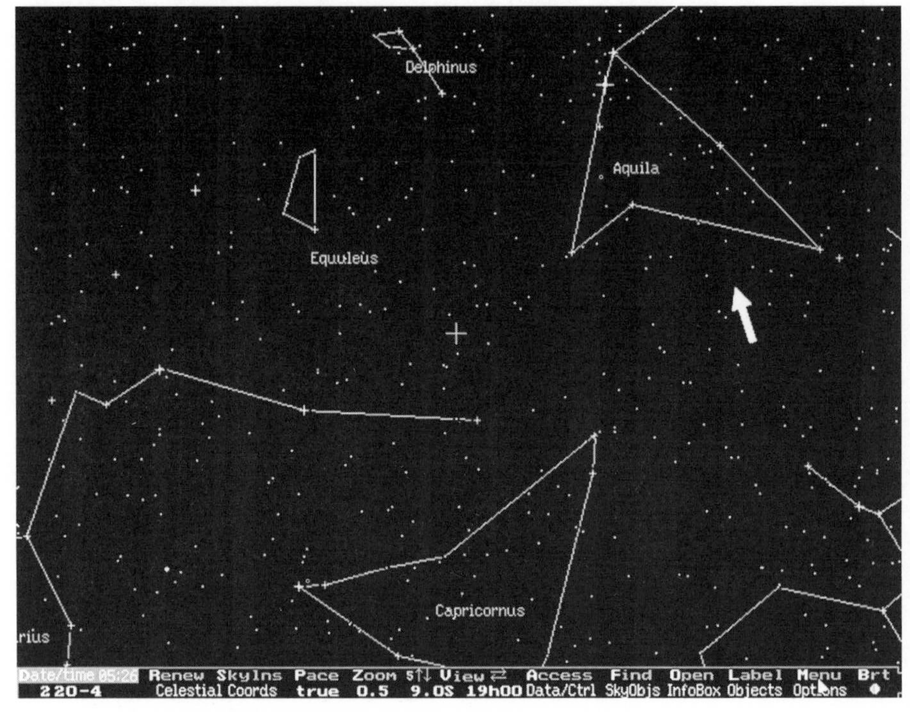

Figure E.1. Nautical twilight, February 20, 5 B.C. DO Aquilae is marked. (Images from *Dance of the Planets,* ARC Science Simulations, www.arcinc.com.)

Other researchers have suggested other stars as possible candidates for the 5 B.C. object observed by the Chinese astronomers, and I will mention these alternatives briefly. As we will see, various researchers have thought along similar lines, but their conclusions differ from mine.

The situation is greatly confused by the separate records of the objects in 5 B.C. from China, and in 4 B.C. from Korea. For this reason, some astronomers have tried to identify the two stars with different objects, while others have regarded them as one and the same. In general, it is fair to say that, prior to the late 1970s, they were universally regarded as different objects and therefore were treated as separate objects by most of the people who investigated them. Today most researchers regard them as being one and the same.

Between 1955 and 1958, the Chinese scientific historian Hsi Tsê-Tsung produced a list of ancient Chinese observations and suggested that there is a radio source close to the position of the 5 B.C. object. This would suggest that the star was most likely a supernova. Two other Chinese scientific historians, Xi Ze-zong and Bo Shu-ren, later pointed out that Hsi was confused about the position of the 5 B.C. and 4 B.C. stars and thus that Xi Ze-zong had made an incorrect identification of the star. We now know that there is no radio source near to where the star appeared, thus the Star of Bethlehem was not a supernova.

In 1965 and 1966, in correcting Hsi's earlier mistake, Xi Ze-zong and Bo Shu-ren made their own identification and suggested that the 4 B.C. star was an outburst of the recurrent nova V500 Aquilae. This nova appeared further west of the suggested position of the 5 B.C. star in the constellation of Aquila. Its galactic coordinates, longitude 047.6 degrees, latitude −09.5 degrees, disagree by a much larger factor with the position of the 5 B.C. star than does DO Aquilae.

V500 Aquilae appeared as a magnitude 6 nova in 1943 (and hence was not well studied because of the Second World War), having brightened from fainter than magnitude 17. B. Kukarin and his Russian colleagues are skeptical about this being the 4 B.C. star, calling it an "unfounded identification"; this rather harsh phrase may partially be due to the rather poor translation of their comments in the mainly Russian-language volume.

In 1969, another Chinese historian of science, T. Kiang, made a new suggestion that again implied that the Chinese star could have been a supernova. He pointed out that it was possible, although with comparatively low probability, that the pulsar PSR 1919 + 10 was formed by the 4 B.C. event. As a pulsar can form only in a supernova explosion, the 4 B.C. star would have to have been a supernova. But this pulsar is a long way from the most probable position of the 5 B.C. star (its galactic coordinates are longitude 047 degrees, latitude −3.9 degrees), and, for this reason, Kukarin and his colleagues also call this an "inacceptable identification" (*sic*).[1] Certainly, if Kiang were correct, the position of the Chinese star given by Richard Stephenson and his colleagues would have to be tremendously in error.

So, as none of the previously suggested identifications are very convincing, let us return to the original question: Can we positively identify the star seen by the Chinese in 5 B.C. with any star known to astronomers today?

The main difficulty is the uncertainty in the position of the 5 B.C. star. If we had an accurate position for this star, the task would be comparatively trivial. Unfortunately, the best that we can do is to estimate its position within a circle some 10 degrees across. An error this big covers a lot of sky; this means that we are probably doomed to failure, however hard we try.

There is, at least in theory, a further and definitive way of deciding the question of the identity of the Star of

Bethlehem. We have to assume that the 5 B.C. star was a nova, but although this method is probably beyond our means at the moment because it is too tedious and probably too difficult to accomplish, there is no harm in mentioning it.

A nova explosion sends out a cloud of gas that expands at anywhere from about 600 to 3,000 miles per second. If we were to discover such an expanding cloud around an old nova, situated in the right part of the sky and not too far from the Sun, and if—calculating the movement of the cloud backwards in time, as Arthur C. Clarke's Jesuit astronomer did in the story "The Star"—we were to find that it was sent out around two thousand years ago, the case of the Star of Bethlehem genuinely would be closed and our star identified.[2] Such a cloud would now be invisible and the only chance of detecting it would be by its very faint radio emission. It would be a difficult task, with no guarantee of success, but it would be a fascinating experiment to perform.

Until then, we can do nothing more than suggest that perhaps, just perhaps, we *do* know which star of the countless millions in the sky was the Star of Bethlehem. If I am right, it may now be called DO Aquilae. It would be quite astonishing if this turns out to be correct. Most likely it is some other, now anonymous star, one that is very faint at the present day, but one that may, even if not for ten thousand years, blaze forth once again in the sky in a renewed outburst.

 APPENDIX 

The Heavens above Bethlehem

IT IS EASY to calculate the exact position of the stars and planets in the sky for any date in the past. Therefore, even though we have no written knowledge of the Star's exact position and date of appearance, we can find out what the sky would have looked like to the Magi at the time when the Chinese star appeared in the sky, using the different possible dates of the event.

Given that Persia, Babylon, and Jerusalem are at almost the same latitude, the sky would have been almost identical when viewed from these three places. It does not matter whether we assume the Magi were Persians or Babylonians, it makes no difference whether they started their odyssey from Tehran or even from Jerusalem; only the local time of twilight would change. Suppose, just for the sake of argument, that the Magi were observing the sky from Tehran in Persia in 5 B.C. What would they have seen at the time that the Star of Bethlehem appeared?

Because we don't know the exact date when the Chinese star appeared, we have to examine several possibilities. Let's take the two dates that straddle the most probable time at which the Chinese star was first seen: February 20 and March 10, 5 B.C.

February 20

Nautical twilight would have occurred at 5:51 A.M. local time. At nautical twilight the Sun is 12 degrees below the horizon and the brightness of the horizon becomes quite considerable, which means that only fairly bright stars can be seen toward the bright red glow in the east. The zenith would still be quite dark, and much of the sky would still be heavily smattered with stars that resist the encroaching dawn (see fig. A.1).

The Moon has just gone down. It is 11.8 days old, 90 percent illuminated, and it set at 5:10 A.M. At nautical twilight the Moon is 8 degrees below the western horizon. Mercury rose in the east at 5:37 and is just 2 degrees high at nautical twilight exactly in the direction of sunrise; it is bright, of magnitude −0.1, and can possibly be glimpsed despite its low altitude in the sky. Mars, bright and red, dominates the south. It is positioned in the head of the Scorpion and, 42 degrees high with magnitude −0.2, is unmistakable; it is close to but brighter than the star Antares (whose name means "the rival of Mars" because of its similar red color). Together, these two bright red stars would have formed a striking pairing in the sky.

Hercules is in the zenith. The fifth largest constellation, it is neither one of the brightest nor the most spectacular. It is straddled by two brilliant stars: Vega, the fifth-brightest star in the sky, leader of the constellation of Lyra the Lyre, is slightly to the east of the zenith; Arcturus, the fourth-brightest star in the sky and the brightest in the constellation of Boötes, is just past the zenith in the west.

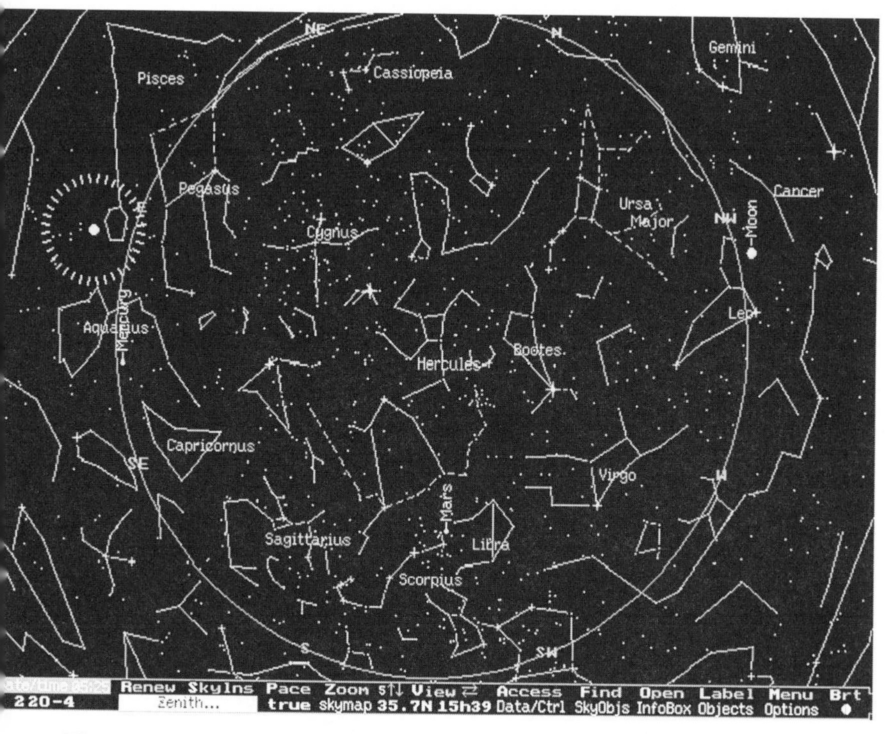

Figure A.1. Nautical twilight on February 20, 5 B.C. (Images from *Dance of the Planets,* ARC Science Simulations, www.arcinc.com.)

Leo the Lion is just setting in the west, straddling the horizon, and Virgo (the Virgin) is dropping down toward the southwestern horizon, although still higher than Leo. In the northeast, Ursa Major, the Great Bear, is sinking, although it will never quite set. As Ursa Major falls, Cassiopeia—the mythical Queen of Ethiopia—is climbing in the northeast.

The south of the sky is dominated by the constellation of Scorpius the Scorpion. From Persia, Babylon, and Jerusalem the whole of Scorpius is visible, including the stars of the sting, which are extremely low or invisible from much of Europe and the northern United States. In the southeast, Sagittarius and the center of the Milky Way dominate.

Below Vega, in the east, Cygnus the Swan is halfway to the zenith. This constellation is also known as the Northern Cross, with the cross-spar being the wings of the swan and the long arm of the cross signifying the neck. The Milky Way extends through Scorpius and Sagittarius in the south and southeast, crossing Cygnus in the east and passing down toward the northeastern horizon as it crosses Cassiopeia.

The region where the Star appeared is low in the east or southeast. Capricornus is just above the southeastern horizon. Aquila the Eagle is rather higher, just below the head of the Swan. Alpha Capricorni (also known as Al Geidi) is 27 degrees high and just around the limit of the twilight glow. Theta Aquilae is no less than 35 degrees high in the east-southeast.

If the Chinese star indeed appeared near Theta Aquilae, it would have been moderately high in the sky at dawn and well away from what would normally be termed a heliacal rising. At nautical twilight it would have been very close to the glow of the impending sunrise but easily visible. Even if it were much farther to the south than we are supposing, close to Alpha Capricorni, it would still be well above the horizon. The difference in the time of rising between the two positions is exactly 29 minutes, so the visibility of the Star would be little different no matter which of the two positions it appeared in.

To obtain a heliacal rising, in the sense that it was just above the horizon at nautical twilight, the Star would have to have been in the southernmost extreme of Capricornus, almost in the constellation of Microscopium. This is far to the south of the supposed position given in the Chinese records and a most unlikely place for a nova to appear, although not impossible for a comet. For the Star to have a had a heliacal rising and still be positioned in the south of Aquila, it would have had to have been really quite faint, for only a rather faint star would have mingled with the glow of twilight and thus only barely been visible at this hour. Alternatively, the Star might have been hidden by moonlight until the Moon had finally set. Both of these explanations, however, seem to contradict totally the idea of a blazing star, brightening the dawn sky.[1]

This is not necessarily a major problem. The Star, if it were a nova, would have appeared suddenly in the dawn sky. The observers, not having seen it there before, would quite reasonably have *thought* they were witnessing the heliacal rising of a star, even though the object of their attention was far higher in the sky than could logically be expected for a first sighting of a star. In this case, the words "en té anatolé" are only saying that the star was first observed in the dawn sky, which is backed up by the alternative translation "in the first light of dawn" (see chapter 2).

At this time the Sun is in Pisces, just below the asterism known as the Circlet. The distance between the Sun and Theta Aquilae is a little more than 50 degrees. This is a quite respectable distance, one that allows an object in this position to rise several hours before the Sun, although only very bright stars will be high enough in the sky to be spectacular before dawn begins to break.

March 10

On this occasion nautical twilight would have occurred at 5:29 A.M. local time. Although many of the stars and constellations are the same or similar to those visible on the previous date, other things have changed completely (see fig. A.2).

The Moon is nowhere to be seen, for it is two days old, just past New, and visible briefly after sunset and consequently completely invisible at sunrise. Mercury has disappeared from the sky as well. It is far too close to the Sun to be seen and rises less than half an hour before sunrise. Only Mars is constant. It is even brighter now, magnitude −0.8, and dominates the south even more than previously. It is still positioned in the head of the Scorpion and, 36 degrees high, remains totally unmistakable. This juxtaposition of Mars and the Rival of Mars (Antares) in the sky would undoubtedly excite the astrologers greatly. They would, no doubt, use the event to provide gloomy predictions of war, strife, and doom to their masters, some of whom would be tottering on their thrones, others keen to cause the tottering of their rivals.

Now, Vega and Lyra occupy the position of honor in the zenith. Cygnus and Deneb (whose name means "tail," reflecting its position in the tail of Cygnus) are high in the east and close to the zenith. Hercules is on the other side of the zenith while Arcturus and Boötes have slid down halfway toward the western horizon.

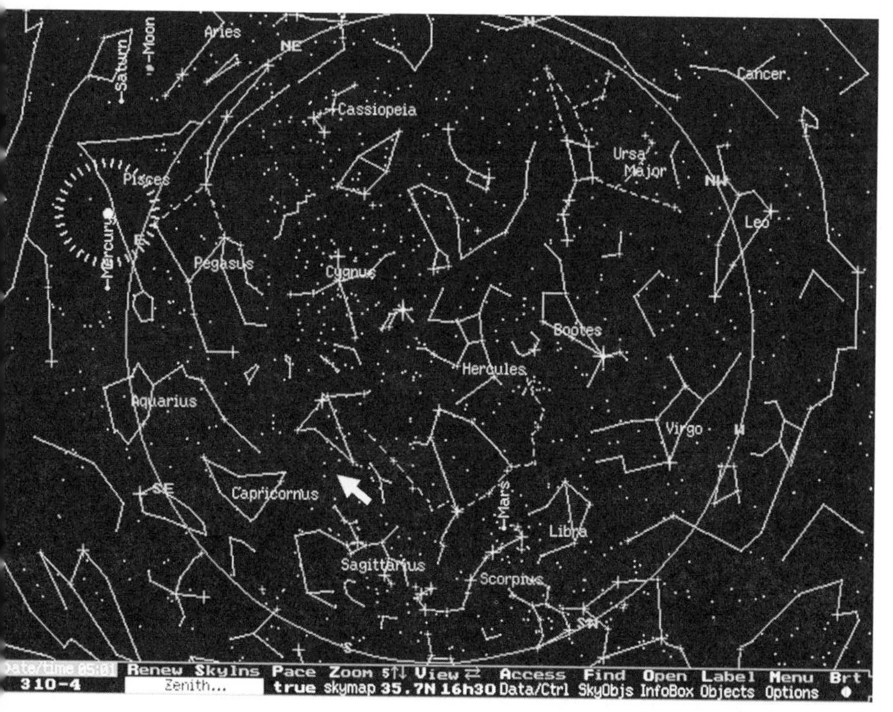

Figure A.2. Nautical twilight on March 10, 5 B.C. (Images from *Dance of the Planets,* ARC Science Simulations, www.arcinc.com.)

Leo has set in the West and is now virtually invisible. Just a few stars of its hindquarters still poke above the horizon. Virgo is only just above the southwestern horizon and about to set. In the northeast, Ursa Major is getting quite low. Cassiopeia is climbing in the northeast, at the same altitude as Ursa Major.

Sagittarius is now in the south and, with it are the gorgeous, misty star clouds of the Galactic Center. Scorpius stretches along, parallel to the horizon, toward the southwest.

In the east, Pegasus has risen, and most of Aquarius can be seen in the southwest. Andromeda can be glimpsed in the dawn glow, but not Pisces. Although a good part of this last constellation is above the horizon, it is composed of quite faint stars that are being defeated by the growing twilight.

The region where the Star appeared is in the southeast and now considerably higher at the start of nautical twilight. Capricornus is farther above the southeastern horizon. Aquila is now quite high in the east-southeast. Alpha Capricorni is now 34 degrees high and above the limit of the bright twilight glow. Theta Aquilae is halfway to the zenith, 44 degrees high.

If the Star appeared around March 10, 5 B.C., close to Theta Aquilae, it would have been high in the sky at dawn. Even if it had appeared in Capricornus, close to Al Geidi, it would still have been at a respectable altitude. It is quite unlikely, even assuming the latter position, that it could be described as having been "low in the dawn sky."

At this time the Sun has entered the constellation of Aquarius. The distance between the Sun and Theta Aquilae is some 65 degrees. Even if the object was a comet and not a nova, this is not an impossible position for a bright comet to appear suddenly, although it is an unexpected one, as it is quite a long way from the Sun in the sky.

May 29

The Chinese star was last seen somewhere around this time. As we have seen, it is quite possible that the monsoon impeded its observation by the Chinese beyond this

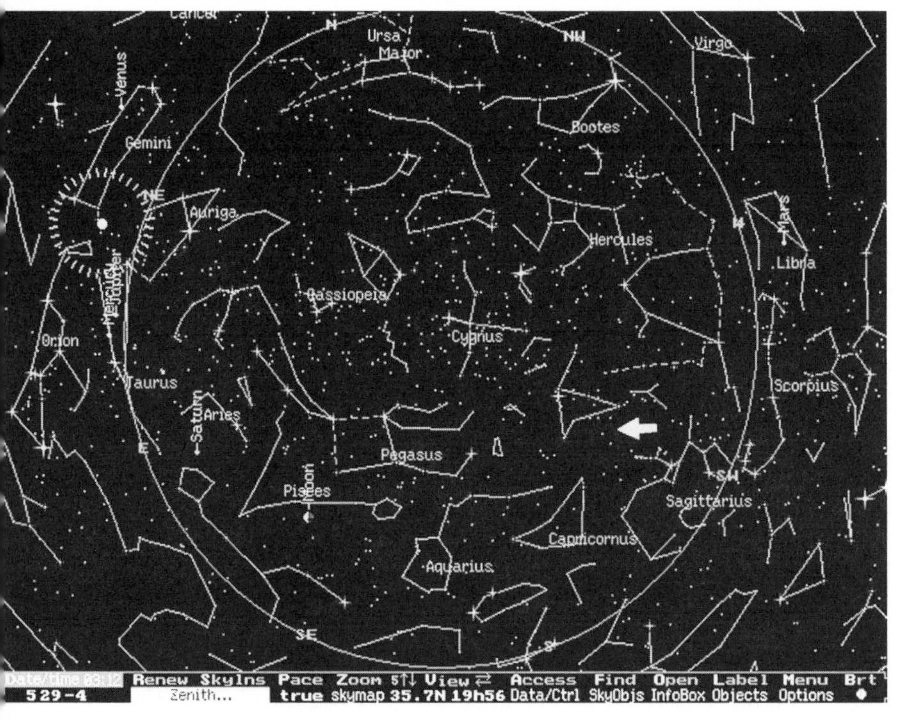

Figure A.3. Nautical twilight on May 29, 5 B.C. The position of the Star is marked with an arrow. (Images from *Dance of the Planets,* ARC Science Simulations, www.arcinc.com.)

date, even though the Star itself would quite possibly still have been clearly visible in a cloudless sky.

Nautical twilight occurs at 3:51 A.M. local time, two hours earlier than in mid-February (see fig. A.3). There is a 40 percent illuminated, waning crescent moon in Pisces, 31 degrees above the eastern horizon. The moon is 23.1 days old and had already risen at 01:09 A.M. Saturn is the only bright planet visible. It is 16 degrees high, above the eastern horizon, in Aries. Jupiter will rise in 10 minutes, at 4:01, and will only be visible in the bright dawn sky. Mercury, much fainter and slightly closer to the Sun, rises at 4:09 and is virtually impossible to see. Both are in Taurus the Bull.

The Star is now in the west-southwest. The Bible indicates that the Star was in the south when the Magi arrived at Bethlehem, so we can work out just when the Star would have been in this position. It turns out that this would have happened in early May, but the Star would have been virtually due south at dawn by mid-April. Theta Aquilae is now 46 degrees high and Al Geidi is 35 degrees. Both would have risen in the east before the last glow of dusk left the opposite horizon, but the sky is still very dark here in the west, so both of them would have been visible all through the night: Theta Aquilae rose just 20 minutes after nautical twilight ended.

Much of the south and east of the sky is filled with large, sprawling, unspectacular constellations such as Cetus (the Whale), Pisces, Aquarius, and Sculptor, with their dreadful dearth of bright stars.

The Milky Way crosses the sky from the southwest, where Sagittarius approaches the horizon, through Cygnus and the zenith, to Cassiopeia, down to Auriga, straddling the northeastern horizon. Ursa Major is at its lowest in the north and bumping along the horizon. Arcturus is just setting in the northeast and will disappear in a few minutes' time, but Vega is still high up in the west, and Hercules, although it is dropping in the sky, is halfway to the zenith.

There is no particular reason for the Chinese to have lost sight of the Star at this time, unless it simply faded from view. The Moon was waning and its light was hardly going to affect observations of the Star for at least the next two weeks. The Star's position, ever more toward the evening sky, was, if anything, becoming ever more favorable for its continued observation. Now we can only speculate that perhaps the monsoon season intervened, so it is possible that, what with the rains and the Moon, many nights passed before there was another clear sky. The time may have been long enough for the Star either to fade slowly from sight before favorable conditions for viewing returned, or to have become so faint that it was no longer recognizable as a bright star. Unfortunately, we will probably never know for certain what exactly did happen.

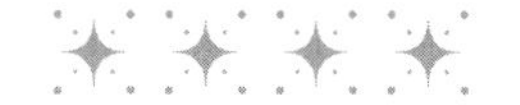

NOTES

Chapter 1. Matthew's Star

1. The New Revised Standard Version does not use the familiar phrase "in the east," found in many other translations of the Bible, although it suggests it as an alternative translation to the one given here.

2. The word used in the original Greek contradicts the version seen in most Nativity plays of a newborn baby Jesus being visited by the Magi. The phrasing suggests that Jesus might have been several months old at the time of their visit.

Chapter 2. Star over Bethlehem?

1. "Εν τη 'ανατολη" in the original Greek.
2. "Εν ται 'ανατολαι" in the original Greek.

Chapter 3. The First Christmas

1. Dionysius wrongly gave the date of the incarnation as 753 A.U.C.

2. Dennis C. Duling and Norman Perrin, *The New Testament: Proclamation and Paranesis, Myth and History* (3d ed.) (New York: Harcourt Brace Jovanovich, 1997).

3. This is so close to the date which is assumed within this text (in day, if not in year) that one wonders if Hippolytus could conceivably have had access to some source of accurate information that has since been lost. The skeptical view is that if enough people make guesses, someone is bound to pick the right date in the end.

4. Because Dionysius Exiguus "forgot" to add the year 0, the millennium is not on December 31, 1999, as only 1,999 years will have passed since January I, A.D. 1, and the start of the Christian calendar. The correct millennium date is December 31, 2000, the end of the 2,000th year A.D. However, this confusion will probably not deter people from celebrating the millennium, as Sir Arthur C. Clarke

once said, "as soon as the three zeros roll up" and, without a doubt, again the following year, when the millennium arrives officially.

Chapter 4. Halley's Comet

1. Please note that the modern constellation boundaries were only fixed in 1932, by a decision of the International Astronomical Union, so it is not necessarily true that, for the Magi, the planets *were* in two different constellations.

2. Paradoxically, the first quarter Moon occurs when the Moon is exactly halfway to full and thus half illuminated. The reason for this is that it has, at that point, completed exactly one quarter of its cycle of phases from New Moon to New Moon.

3. Of the 102 occultations listed in table 3.3, as we can see, nine were theoretically visible from Babylon, but, of these, just two, both of Mars, actually happened at night and would thus have been visible to the Magi. Although Mars was very bright on both occasions (around magnitude -1), both occurred close to Full Moon and would thus have had more limited impact.

Chapter 5. Shooting Stars

1. Strictly speaking, Comet Halley is as old as the solar system (nearly five thousand million years old); the age given here refers to the time that the comet has spent in the inner solar system, visible from the Earth.

2. Despite a foul, foggy, freezing London evening, the enchantment of the presentation far surpassed the discomfort of the journey to attend.

3. Arthur Clarke, "The Star."

Chapter 6. Supernova Bethlehem?

1. I am indebted to William G. Contento, of the Locus Science Fiction Guide (http:www.sff.net/locus), for providing me with the publication details, which are surprisingly not revealed by Arthur C. Clarke in his notes to the collection where this story was presented.

2. We now know that this affirmation has been superseded. On January 23, 1999, a gamma ray burst (GRB) was detected in the constellation of Corona Borealis. This explosion, of uncertain origin, reached magnitude 9, despite being about 10,000 million light years away. At 10,000,000,000,000,000 times the luminosity of the Sun, it far outshines any supernova.

3. In fact, because our eyes detect red light inefficiently compared to other colors, a red star appears comparatively fainter than an identical blue or white one. If Sir John Herschel observed Betelgeuse to be similar in brightness to the bluish Rigel, it must really have been significantly brighter.

Chapter 7. We Three Kings?

1. In Spanish, *los Reyes de Oriente* or *Sus Majestades, los Reyes de Oriente* (Their Majesties, the Kings of Orient.)

2. In the Spanish tradition the kings leave presents for good children and coal for bad ones. This practice has led to the popularization of the sale of sweet coal, made from sugar, at Christmas.

3. A highly comprehensive account of how the image of the Magi has changed over the centuries to the present day is given by historian Richard Trexler of the State University of New York at Binghamton in his book *The Journey of the Magi—Meanings in History of a Christian Story,* which is the source for some of the material presented here.

4. Originally a tribute to King David, it appears that the Psalm as a whole goes beyond David to the coming Messiah.

5. The Reverend Phillip Greetham is an English Methodist minister, currently at Finton-on-Sea, in Essex (United Kingdom), who has wide-ranging interests including Christianity and astronomy. His Internet site on the Star of Bethlehem is by far the best I have seen and presents an enormous amount of historical and biblical information from a Christian, but also scientific viewpoint, with no strong prejudices or preconceived opinions. I strongly recommend readers to check this site at "http://ourworld.compuserve.com/homepages/p_greetham/Wisemen" to get a fascinating insight into the Nativity and the events surrounding it.

6. In the Catalan regions of Spain (Catalonia, Valencia, Majorca, and also in the state of Andorra), "Joan" is a common boy's name; it is the Catalan spelling of Juan, the Spanish version of the name John. There are various spellings of the name of Balthasar. Like the name of Shakespeare, which the playwright himself spelled as many as thirteen different ways, the spelling of these names was only completely standardized later.

7. Such an account would seem plausible because Ethiopia has one of the oldest surviving Christian churches in the world, a discovery that astonished and disconcerted missionaries dispatched to the country during the last century to "convert the natives." Furthermore, despite the country's long isolation, it also had a significant Jewish colony, leading to the famous and spectacular rescue of the Sephardic community carried out by the Israeli government a few years ago. Although it is not obvious how contact came about initially, Ethiopia evidently had important ties with the Jewish and early Christian churches at some point in the distant past. It would have seemed plausible that an African magus (singular of "magi") could have been an Ethiopian.

8. The recipe for holy anointing oil is given in Exodus 30:34–36:

> 34 The Lord said to Moses: Take sweet spices, stacte, and onycha, and galbanum, sweet (spices with pure frankincense (an equal part of each),
>
> 35 and make an incense blended as by the purfumer, seasoned with salt, pure and holy;
>
> 36 and you shall beat some of it into powder, and put part of it before the covenant in the tent of meeting where I shall meet you; it shall be for you most holy.

9. *Nature* 264 (1976): 513–517.

10. *Quarterly Journal of the Royal Astronomical Society* 32 (1991).

11. Daniel 2:2.

12. Daniel 2:24.

13. An alternative translation of this phrase, given in other versions of the Bible, is "perverting the ways of the Lord," which is even stronger. See also Acts 13:10.

14. Much of the Old Testament book of Daniel deals with King Nebuchadnezzar of Babylon and his relationship with the Jews after

the fall of Jerusalem, showing that the Babylonians were steeped in Jewish tradition and prophecy.

15. These diaries are not on display to the public, but are available for study (one of the drawers in Walker's office is filled with fragments of tablets, which are stored with care and taken out only when necessary). One particularly interesting anecdote that illustrates some of the problems associated with the Babylonian tablets is the fact that these are, of course, no more than dried mud: I have it on good authority that one American museum's collection of tablets was stored in a basement, supposedly completely safe, until a disastrous flood left the basement inundated and the tablets as small globs of mud on the floor.

16. The tablet is classified as BM 35429.

17. The similarity between these two stories leads me to suspect that one is a corruption of the other, with names and places changed, although I cannot prove this.

Chapter 8. Triple Conjunctions

1. Pluto, as usual, is an exception, for its orbit is so highly inclined that it will rarely show conjunctions as such with other planets. However, as Pluto is also around seven magnitudes fainter than Neptune, conjunctions of Pluto would be of scarce interest and even scarcer visibility.

2. Recently I had the chance to see just how spectacular this configuration would have been. At the end of a night's observing, I went out to look at the dawn and was struck by a beautiful pair of bright "stars" low on the horizon: Jupiter and Mercury. Even though sunrise was imminent, the two were extremely striking. A four-planet configuration would have been even more stunning.

Chapter 9. Is the Answer in Chinese?

1. Because of the "0.3" of a day (8 hours), not all eclipses observed in one place are followed by another a Saros later: sometimes the eight-hour shift moves the following eclipse to nighttime when it is, of course, invisible.

2. This distance is the same as that between the two "pointer" stars of the Plough, or Big Dipper, that point the way to the Pole Star.

3. Another nova, Nova Ophiuchi of 1848, was also at quite high galactic latitude (18 degrees). Some references state that this nova was also possibly rather bright (magnitude 2), others that it was quite faint (magnitude 4). The bright Nova Coronae of 1866 was at galactic latitude +48 degrees.

Chapter 10. What Was the Star?

1. If, as a few experts have suggested, the Magi came from Arabia and not from Babylon or Persia, their journey would have been very simple. From Arabia they would have traveled around the coast by boat, disembarking at the head of the Red Sea. From there, there was a well-transited trading route called the King's Highway, which would rapidly carry them the relatively short distance north to Jerusalem. There would be no desert to cross. By boat and on horseback they would have made the journey in a couple of weeks at most, and possibly in just a week or so.

2. I once rode a camel on the Canary Island of Lanzarote. In the Timanfaya National Park, visitors can mount a camel and take a ride across a stretch of the local desert sand to a high point nearby, one that offers a good view of the volcanoes of the park. They plod along pretty slowly but very steadily up the slope, led by a handler. It's quite an experience. Supposedly, when hard pressed, a camel can do 60 miles per day for a short time period.

Epilogue

1. They suggest that the 5 B.C. star was Venus (which, as we have seen, is rather unlikely) and that the 4 B.C. star was a comet.

2. It would be wonderful to vindicate Arthur C. Clarke in this way, particularly as he has been successful with so many other (presumably) tongue-in-cheek predictions in his stories and novels.

Appendix

1. Assuming, of course, that the Star was very bright, which is not proven, though it seems likely from Chinese observation records.

 BIBLIOGRAPHY 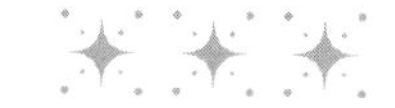

Chapter 1. Matthew's Star

Note: Biblical quotations are taken from the New Revised Standard Version of the Holy Bible (Anglicized edition), published by Oxford University Press (1995). Other biblical material is taken from the text and from the extensive notes accompanying the new version of the translation of this Bible into Spanish (edited by Friar Serafín de Ausejo) of the original Greek texts (2d ed., 1986), published by Editorial Herder; this Spanish Bible is translated from the Nestlé-Aland version of the Greek text and from the Greek New Testament, which is the modern Greek version of the Biblical Societies of Europe and America.

Much historical material is taken from the Greek New Testament along with the first three sources cited below; the beautiful article by Sinnot; the text by Neirynck; and the extensive marginal notes of an anonymous biblical scholar who read the first version of this chapter for Princeton University Press.

Hertz, J. H. *The Pentateuch and Halftorahs: Hebrew Text, English Translation, and Commentary*. 2d ed. London: Soncino Press, 1976.

Hughes, D. W. "The Star of Bethlehem." *Nature* **264**, 513 (1976).

Humphreys, C. J. "The Star of Bethlehem—A comet in 5 B.C.—and the date of the birth of Christ." *Quarterly Journal of the Royal Astronomical Society* **32**, 389 (1991).

Kidger, M. R. "Una Estrella Sobre Belén." *Tribuna de Astronomía* **69**, 32 (1992).

Kukarin, B. V., Kholopov, P. N., Pskovsky, Yu. P., Efremov, Yu. N., Kukarina, N. P., Kurochkin, N. E., Medvedeva, G. I., Perova, N. B., Fedorovich, V. P., and Frolov, M. S. *General Catalogue of Variable Stars*. 3d ed. Moscow, 1972.

Molnar, M. R. "The Magi's Star from the perspective of ancient astrological practices." *Quarterly Journal of the Royal Astronomical Society* **36**, 109 (1995).

Moore, P. A. "The stars from the bottom of a well." In *Patrick Moore's Armchair Astronomy*. Wellingborough, U.K.: Patrick Stephens, 1984.

Neirynck, F. "The Synoptic Problem." In *The New Jerome Biblical Commentary*. Ed. Raymond E. Brown. New York: Prentice Hall, 1989.

Paffenroth, K. "The Star of Bethlehem casts light on its modern interpreters." *Quarterly Journal of the Royal Astronomical Society* **34**, 449 (1993).

Perrin, N., and Duling, D. D. *The New Testament: An Introduction*. 2d ed. New York: Harcourt Brace Jovanovich, 1982.

Radice, B. *The Letters of the Younger Pliny*. London: Penguin Books, 1963.

Sinnott, R. W. "Thoughts on the Star of Bethlehem." *Sky and Telescope* **36**, 384 (1968).

Chapter 3. The First Christmas

Bulmer-Thomas, I. "The Star of Bethlehem." *Quarterly Journal of the Royal Astronomical Society* **33**, 363 (1992).

Clark, D. H., Parkinson, H., and Stephenson, F. R. "An astronomical re-appraisal of the Star of Bethlehem—a nova in 5 B.C." *Quarterly Journal of the Royal Astronomical Society* **18**, 443 (1977).

Finegan, J. *Handbook of Biblical Chronology*. Princeton: Princeton University Press, 1964.

Hughes, D. W. "The Star of Bethlehem." *Nature* **264**, 513 (1976).

Humphreys, C. J. "The Star of Bethlehem—a comet in 5 B.C.—and the date of the birth of Christ." *Quarterly Journal of the Royal Astronomical Society* **32**, 389 (1991).

Keller, W. *Bible as History*. 2d rev. ed. New York: Morrow, 1981.

Chapter 4. Halley's Comet

Chown, M. "O invisible Star of Bethlehem." *New Scientist*, December 23/30, 1995, pp. 34–35.

Gore, J. E. *The Scenary of the Heavens: A Popular Account of Astronomical Wonders*. London: Roper and Drowley, 1890.

Hughes, D. W., Yau, K.K.C., and Stephenson F. R., "Giotto's

Comet—was it the Comet of 1304 and not Comet Halley?" *Quarterly Journal of the Royal Astronomical Society* **34**, 21 (1993).

Kidger, M. R. 1994, 1995. "La estrella de Belén." Christmas Lecture, Museo de las Ciencias y del Cosmos, La Laguna (Tenerife, Spain).

Molnar, Michael R. "The coins of Antioch." *Sky and Telescope*, January 1992, pp. 37–39.

Molnar, Michael R. "The Magi's Star from the perspective of ancient astrological practices." *Quarterly Journal of the Royal Astronomical Society* **36**, 109 (1995).

Moore, P. A. "The Star of Bethlehem." London Planetarium, Christmas Lecture, 1991.

Olson, R.J.M. "Much Ado about Giotto's Comet." *Quarterly Journal of the Royal Astronomical Society* **35**, 145 (1994).

Olson, R.J.M., and Pasachoff, J. M. "New information on Comet Halley as depicted by Giotto di Bondone and other Western artists." *Proceedings of the 20th ESLAB Symposium on the Exploration of Halley's Comet* **3**, 201, ESA SP-250 (1986).

Olson, R.J.M., and Pasachoff, J. M. "New information on Comet Halley as depicted by Giotto di Bondone and other Western artists." *Astronomy and Astrophysics* **187**, 1 (1987).

Rada, W. S., and Stephenson, F. S. *Quarterly Journal of the Royal Astronomical Society* **34**, 331 (1992).

Sinnot, R. W. "Thoughts on the Star of Bethlehem." *Sky and Telescope* **36**, 384 (1968).

Various. 1987. "The Astronomy Encyclopaedia." General Editor, Patrick Moore. London: Michell Beazley.

Yeomans, D. K. *Comets: A Chronological History of Observation, Science, Myth, and Folklore.* New York: John Wiley, 1991.

Yeomans, D. K., Rahe, J., and Fretag, R. 1986. "Halley's Comet in history." *Space Missions to Halley's Comet*, ESA SP-1066 (1986).

Chapter 5. Shooting Stars

Hughes, D. W. "The Star of Bethlehem." *Nature* **264**, 513 (1976).

Jenniskens, P. "Meteor stream activity." *Astronomy and Astrophysics* **295**, 206 (1995).

Kidger, M. R. 1994, 1995. "La estrella de Belén." Christmas Lecture, Museo de las Ciencias y del Cosmos, La Laguna (Tenerife, Spain).

Moore, P. A. "The Star of Bethlehem." London Planetarium, Christmas Lecture, 1991.

Chapter 6. Supernova Bethlehem?

Clark, D. H., and Stephenson, F. R. *The Historical Supernovae.* Oxford: Pergamon, 1977.

Clarke, Arthur C. "The Star of the Magi" (1954). Published in the collection *Report on Planet Three and Other Speculations.* London: Corgi Books, 1975.

Clarke, Arthur C. "The Star." (1955). Published in the collection *The Other Side of the Sky.* New York: Victor Gollancz Science Fiction, 1987.

Hsü, K. J. *The Great Dying: Cosmic Catastrophe, Dinosaurs and the Theory of Evolution.* 1986; Spanish translation: Barcelona: Antoni Bosch Editor, 1993.

Kippenhahn, R. *100 Billion Suns.* Paperback edition. Princeton: Princeton University Press, 1993.

Marschall, L. A. *The Supernova Story.* Paperback edition. Princeton: Princeton University Press, 1994.

Mitton, S. *The Crab Nebula.* London: Faber and Faber, 1979.

Moore, P. A. *The Guinness Book of Astronomy Facts and Feats.* 1st ed. Enfield, U.K.: Guinness Superlatives, 1979.

Murdin, P., and Murdin, L. *Supernovae.* Cambridge, U.K.: Cambridge University Press, 1985.

Niven, L., and Pournelle, J. *The Mote in God's Eye.* London: Donald Futura edition, 1983.

Niven, L., and Pournelle, J. *The Gripping Hand.* New York: Pocket Books, 1994.

Saha, A., Labhardt, L., Schwengeler, H., Macchetto, F. D., Panagia, N., and Tammann, G. A. "Discovery of Cepheids in IC 4182: Absolute peak brightness of SNIa 1937C and the value of H_0." *Astrophysical Journal* **425**, 14 (1994).

Sandage, A. "Astronomical problems for the next three decades." In *Key Problems in Astronomy,* ed. G. Münch, A. Mampaso, and F. Sánchez. Cambridge, U.K.: Cambridge University Press, 1997.

Webb, T. J. *Celestial Objects for Common Telescopes*, vol. 2: The Stars. Rev. and enlarged republication of original 6th ed. published in 1917. New York: Dover, 1962.

Chapter 7. We Three Kings

Antoniadi, E. M. *La planéte Mars, 1659–1929*. Paris: Librarie Scientifique Hermann et Cie, 1930. (English translation by Patrick Moore. Shaldon, U.K.: Keith Reid Ltd., 1975.)

Antoniadi, E. M. *La planéte Mercure*. Paris: Gauthier-Villars, 1934. (English translation by Patrick Moore. Shaldon, U.K.: Keith Reid, 1974.)

Clark, D. H., and Stephenson, F. R. *The Historical Supernovae*. Oxford: Pergamon, 1977.

Clark, D. H., Parkinson, J. H., and Stephenson, F. R. "An astronomical re-appraisal of the Star of Bethlehem—a nova in 5 B.C." *Quarterly Journal of the Royal Astronomical Society* **18**, 443 (1977).

Hunt, G. E., and Moore, P. A. *The Planet Venus*. London: Faber and Faber, 1982.

Martos-Rubio, A. *Historia de las constelaciones: Un ensayo sobre su origen*. Madrid: Equipo Sirius, 1992.

Moore, P. A. *Guide to Mars*. Guilford, U.K.: Butterworth Press, 1977.

Moore, P. A. *The Guinness Book of Astronomy Facts and Feats*. Enfield, U.K.: Guinnes Superlatives, 1979.

Norton, A. P. *Norton's 2000.0—Star Atlas and Reference Handbook*. 18th ed. Ed. I. Ridpath. Harlow, U.K.: Longman Scientific and Technical, 1989.

Roaf, M. *Mesopotamia and the Ancient Middle East*. Vol. 1. Spanish translation. Madrid: Ediciones Folio/Ediciones del Prado, 1992.

Roaf, M. *Mesopotamia and the Ancient Middle East*. Vol. 2. Spanish translation. Madrid: Ediciones Folio/Ediciones del Prado, 1992.

Sachs, A. "Babylonian horoscopes." *Journal of Cuneiform Studies* **6**, 49 (1952).

Sachs, A. J., and Walker, C.B.F. "Kepler's view of the Star of Bethlehem and the Babylonian almanac for 7/6 B.C." *Iraq* **46**, 43 (1984).

Stephenson, F. R. "A revised catalogue of pre-telescopic Galactic novae and supernovae." *QJRAS* **17**, 121 (1976).

Sinnott, R. W. "Thoughts on the Star of Bethlehem." *Sky and Telescope* **36**, 384 (1968).

Strom, R. G. *Mercury, the Elusive Planet.* Washington, D.C.: Smithsonian Institution Press, 1987.

Swerdlow, N. M. *Babylonian Theory of the Planets.* Princeton: Princeton University Press, 1997.

Trexler, R. C. *The Journey of the Magi: Meanings in History of a Christian Story.* Princeton: Princeton University Press, 1997.

Vogt, W. M., and Gnam, C.A., Jr., eds. "Battlefield of sand." In *Desert Storm.* Leesburg: Empire Press, 1991.

Zirker, J. B. *Total Eclipses of the Sun.* Expanded paperback ed. Princeton: Princeton University Press, 1995.

Epilogue

Hsi Tzê-Tseung. "A new catalogue of ancient novae." *Acta Astronomica Sinica* **3**, 183 (1955).

Hsi Tzê-Tseung. "A new catalog of ancient novae." *Smithsonian Magazine* **2**, 109 (1958).

Kiang, T. "Possible dates of birth of pulsars from Ancient Chinese records." *Nature* **223**, 599 (1969).

Kukarin, B. V., Kholopov, P. N., Pskovsky, Yu. P., Efremov, Yu. N., Kukarina, N. P., Kurochkin, N. E., Medvedeva, G. I., Perova, N. B., Fedorovich, V. P., and Frolov, M. S. *General Catalogue of Variable Stars.* 3d ed. Moscow, 1971.

Xi Ze-zong, and Bo Shu-ren. "Ancient novae and supernovae recorded in Chinese, Korean and Japanese annals and their significance in radioastronomy." *Acta Astronomica Sinica* **13**, no. 6 (1965).

Xi Ze-zong, and Bo Shu-ren. "Ancient oriental records of novae and supernovae." *Science* **154**, 597 (1966).

INDEX

Acts of the Apostles, 9
Adoration of the Magi, 83–84, 85
Against Heresies, 8
Agrippa, Marcus, 94
Ahura-Mazda, 192–93
Altair, 235, 237, 238
Antioch coin, 104–9
Aquila constellation, 235, 239, 242, 245, 259, 267, 268–69. *See also* Theta Aquilae
Arabia, 195, 294n.1
Arabic astronomical chronicles, 220, 223
Aramaic, 7, 10
Archelaus, 49
Ashurbanipal, Library of, 181–82
Assyria, 180
Astronomical event, star as, 20, 23–38
Astronomy of Babylonia, 183–92
Astronomy of the Bible, The, 36
Athanasius, 17
Augustine of Hippo, 170
Augustus Caesar, 45, 53; census of, 12, 24, 52–53, 54–57

Babylonia, 178–92, 196–97, 261, 292–3n.15; astronomy of, 183–92, 230, 231, 232; writings from, 181–83
Babylonian captivity, 10, 178–79, 196–97
Balaam, 13–14, 15, 16, 20, 207, 258
Balaq, King of Moab, 14
Balthasar, 170, 172, 174
Balthezar, Joan, 172, 174
Bar Cozeba, 15
Bar-Jesus, 177
Bayeux Tapestry, 90, 229
Bede, Venerable, 172
ben Halafta, Yose, 43
Betelgeuse, 143–44, 153, 155–57
Bethlehem Supernova, 148. *See also* Nova of 5 B.C.
Bible, New Revised Standard Version, 27, 69, 168, 289n.1
Bible as History, The, 62
Binary star systems, 141–42, 144, 148, 233
Birth of Jesus of Nazareth, 39–68; date of, 57–68; year of, 39–57
Bokenkotter, Thomas, 44
Bo Shu-ren, 273
Brahe, Tycho, 160–61, 162–63, 233, 240–41
Brito, Andrés, 57

Calendar: fixing of modern dates of, 43–45; Gregorian, 41–42; Julian, 41–42; Roman, 40–41
Calixtus III, Pope, 90
Calvary, 3
Capricornus constellation, 235, 236, 242, 244, 245, 259, 263–64, 267
Census of Caesar Augustus, 12, 24, 52–53, 54–57
Chadwick, Henry, 19
Chaldea, 180
Chandrasekhar Limit, 146, 147
Chant, C. A., 124
Ch'ien-han-shu, 234–35
Chinese astronomical chronicles, 164, 219–46; and astronomical appearance at time of Jesus'

Chinese astronomical chronicles (*cont.*)
birth, 234–46; and eclipse of the sun, 220–22; and meteor showers, 224–25; and the nova of 5 B.C., 259–66; and observation of comets, 223, 226–31; and sunspot observation, 223–24
Chown, Marcus, 106
Christian oral tradition, 5
Christmas, 57–60
Chronicle of Zugnin, 170
Clark, David, 233, 237, 239, 241, 269
Clarke, Arthur C., 136–39, 163, 275
Clement of Alexandria, 61, 192, 195
Codex Sinaitacus, 10
Codex Vaticanus, 10
Colossians, 9
Comet, 33, 116, 186, 223, 226–31, 235. *See also* Halley's Comet
Comets—A Descriptive Catalog, 85
Concise History of the Catholic Church, 44
Conjunction of the planets, 26, 95–100, 188–91, 199–218. *See also* Triple conjunctions
Contra Celsum, 18–19
Crab Supernova, 158–60, 233
Cyrilid meteor shower, 120, 123–26

Daniel (Old Testament book of), 176–77, 292–3n.15
David, King, 14, 15
Death of Jesus of Nazareth, 68–72
Degenerate matter, 145, 146
De Pascha Computus, 61
Dionysius Exiguus, 43–45, 52, 57
DO Aquilae, 269–71, 273, 275
Dodd, Robert, 112–13
Duling, Dennis, 46

East as location of star, 25, 26, 33–34, 262, 281, 289n.1
Eclipse of moon and Herod's death, 44–51, 67–68
Eclipse of sun, 184; and the Chinese, 220–22; and Jesus' death, 69–72; predictions of, 185–86
Edward the Confessor, 87
Elymus, 177
Epiphanius, Bishop of Salamis, 61
Epistle to the Ephesians, 17–18
Eta Aquarids, 115, 225
Ethiopia, 172, 174
European Space Agency (ESA), 84
Eusebius, 8

Fast nova, 245–46. *See also* Nova
Finegan, Jack, 57–58, 61
Fireball, 111–14
Frankincense, 174–75

Galactic supernovas, 157–65
Galilei, Galileo, 98, 224
Gaspar, 170, 172
General Catalog of Variable Stars, The, 268
Genesis (Old Testament book of), 190
Giotto di Bondone, 82–84, 85
Godwinson, Harold, 87–90
Gold, 174
Golgotha, 3
Gore, J. E., 80
Gospels, viii, 3–11. *See also* individual Gospels by name
Greetham, Phillip, 99, 291n.6; and the Magi, 171, 176, 177, 179; and Persia, 192, 194
Gregorian calendar, 41–42

Halley, William, 91
Halley's Comet, 74, 82–95, 117, 184, 251; Chinese observation of, 224, 226, 227, 240, 242; and the Nativity, 92–95

Handbook of Bible Chronology, 58
Harris, John, 99
Hebrew, 10
Heliacal rising, 26, 28, 281
Herod, King, 4, 13, 24, 108, 256; character of, 32, 63; death of, 46–49; and the Magi, 12, 17, 32–33, 260, 261–62; and sighting of the Star, 19, 29; and the slaughter of the innocents, 51–52, 63
Herodotus, 176
Herrick, Edward C., 110
Herschel, John, 155–56
Hind, John, 94
Hippolytus, 61
Historical event, Star as, 23–38
History of the Early Church, 8
History of Three Kingdoms—the Chronicle of Silla (Samguk Sagi), 235
Homestake Gold Mine, 37
Hsi Tsê-Tsung, 273
Hubble Space Telescope, 147, 155
Hughes, David, 19, 23, 63, 84, 239–40; and article in *Nature*, 25–27, 66–67, 121; and chronology of the nativity, 66–67; and location of Star in sky, 25–27, 35; and Magi, 175–76, 193; and meteors, 121, 126
Humphreys, Colin, 56, 64–67, 176, 179, 196
Hunt, Garry, 79
Hyakutake, Comet, 240, 243

Ibn Butlan, 159
Ideler, Christian Ludwig, 204
Ignatius of Antioch, 10, 17–18, 28–29
Infancy, Gospel of the, 179
Irenaeus, Bishop of Lyons, 6, 8
Isaiah (Old Testament book of), 169, 175, 195
Istar (Ishtar), 76–77, 187–88. *See also* Venus

James, Gospel of, 16, 17
Japanese astronomical chronicles, 165, 220, 230
Jesus of Nazareth, 3, 5, 6, 51; birth date of, 39–68; death date of, 68–72; and the Magi, 12, 280n.2
Jet Propulsion Laboratory, 227
John, Gospel of, 5, 6
Joseph, 12, 13, 52, 53
Josephus, 9
Jupiter, 96–99, 102, 103, 104–8, 116, 129, 130, 188–89; and massing of planets, 256–57; and pairing with Moon, 257, 258; and triple conjunction, 199–218, 254–56
Justin Martyr, 179, 195

Keller, Werner, 62
Kepler, Johannes, vii–viii, 161, 203–4
Kiang, T., 274
Kirch, Gottfried, 130
Korean astronomical chronicles, 164, 220, 230, 235–46
Kronk, Gary, 84–85
Kukarin, B. V., 236–37, 273

Length of visibility, 29–34, 120–22, 126, 134
Leonid meteor storms, 127–30, 131–32, 133–34, 225
Lipscombe, Trevor, 69–70
Location of Star, 25, 26, 33–34, 246. *See also* Nova of 5 B.C.
Luke, Gospel of, 5–7, 8–9; and birth date of Jesus, 52–53, 54, 62–63; and chronology of the nativity, 64–65; and eclipse at death of Jesus, 69, 72; and the

Luke, Gospel of (*cont.*)
Nativity, 11–12; writing of, 9–10
Lyrids, 115, 225–26

Magi, 27–28, 166–97, 289n.2; definition of, 81, 175–78; gifts of, 59, 170, 174–75; and Gospel of James, 17; and Gospel of Luke, 12; and Gospel of Matthew, 4, 12, 29–34, 166–67; journey of, 29–34, 259–66; as kings, 168–70; and meteors, 113–14, 119, 121–22, 125–26; names and origin of, 170–74, 178–80, 193, 294n.1; and occultation of moon, 108–9
Magnitude scale, 154–55
Marco Polo, 193
Mariner II, 78
Mark, Gospel of, 5–7
Mark, John, 8
Mars, 188–89, 201, 257
Martin, Ernest, 66–67
Martos-Rubio, Alberto, 184
Mary, 4, 11–12, 13, 52, 53, 266
Masada, 9
Mass transfer, 144
Matthew, Gospel of, 3–7, 8–9, 13–19, 22, 23, 51; and chronology of the nativity, 64–65; earliest manuscripts of, 10–11; and journey of the Magi, 31–34, 166–67; and location of the Star, 25, 26, 33–35
Maunder, Edward W., 36
Maunder Minimum, 36, 224
Melchior, 170, 172, 174
Mesopotamia, 180
Messiah, 14, 15–16, 169, 207; announced by Star, 13, 14, 15–16, 17, 20, 259; and Regulus, 214; search for, and Babylonia conquest, 214–15; and Zoroastrianism, 192, 193, 194, 196–97
Messier, Charles, 91–92
Meteor, 110–35, 224–25
Meteor storm, 127–35
Millennium, celebration of, 68, 289–90n.4
Miraculous event, Star as, viii, ix, x, 20–21, 22–23
Miramme, 32
Molnar, Michael, 105, 108, 109
Moon: and dimming of Star, 262–64; occultation by, 100–109; and pairing with Jupiter, 257, 258; and supernovas, 139, 148
Moore, Patrick, 37, 81, 118, 120
Moses, 14
Muñoz, Jerome, 241
Myrrh, 175
Myth, Star as, 20, 21–22

Nativity, viii, 11, 15, 64–68, 250; and Gospel of Luke, 11–12; and Gospel of Matthew, 4–5, 12
Nebuchadnezzar, King, 177
Neptune, 98–99, 201
Nicholas, Saint, 167
Nicodemus, 16
Nile, 26–27
Ninevah tablets, 77
Nova, 233–34, 244, 259–66; bright, 265; recurrent, 234, 267–68
Nova of 5 B.C.: and the Chinese astronomical chronicles, 259–66; location of, 246, 259, 280, 284, 286–87. *See also* Supernova Bethlehem
Numbers (Old Testament book of), 13

Occultation, 95, 100–109, 188
Og, King of Bashan, 14

O'Keefe, John, 124
Old Testament, 14–15, 169. *See also* Daniel; Genesis; Isaiah; Numbers; Torah
Origen, 18–19, 171, 172
Orionids, 115, 225

Paez, Pedro, 174
Palitzsch, Johann, 92
Papias, Bishop of Hieropolis, 7–8
Parallax effect, 101, 202
Parkinson, John, 237, 239
Passover, 46–48
Paul, 9, 177
Perrin, Norman, 46
Perseids, 110–11, 115, 118, 132, 225
Persia and the Magi, 179, 192–96, 261
Peter, 7–8
Peter, Gospel of, 69
Pieces: and double pairings, 257–59; and massing of planets, 256–57; and the triple conjunction, 254–56
Plato, 6
Pliny the Younger, 21–22
Polycarp, Letter to, 10
Portents, 21–22
Protoevangelium of James, 16, 17, 18, 19, 28–29
PSR 1919 + 10, 274
Ptolemy, 154

"Q" (Source Gospel), 9
Quadrantids, 114
Quasars, 139
Quirinius, 52, 53, 54

Recurrent nova, 234, 267–68. *See also* Nova
Red giant star, 143–44, 146, 153
Regulus, 214
Roche Lobe, 144, 146

Sargon the Great, 181
Saros cycle, 185, 221, 293n.1
Saturn, 99–100, 188–89; and massing of planets, 256–57; and pairing with Mars, 257; and triple conjunction, 199–218, 255–56
Scenery of the Heavens, The, 80
Scrovegni Chapel, 82, 85, 93
Signs indicating Star of Bethlehem, 247–66; and triple conjunction (first), 254–56; and massing of planets (second), 256; and two pairings in Pisces (third), 257–59; and nova (final), 259–66
Signs indicating Star of Bethlehem, 247–66
Sihon, King of Amorites, 14
Sinnott, Roger, 96–97, 193
Sirius, 26–27, 144–45, 155
Sky at time of Jesus' birth, 277–87; February 20, 278–82; March 10, 282–84; May, 2, 284–87
Slaughter of the innocents, 51–52, 63
Sol Invictus, festival of, 58–60
Sosigenes, 41
Spectra, 140
"Star, The," 136–37, 275
"Star of the Magi, The," 137
Stephenson, Richard, 184, 233, 237, 239, 241, 269
Sternberg Institute, 236
Steward Observatory, 47
Stolovy, Susan, 47
Sumeria, 180–81
Sunspot observation, 36, 223
Supernova, 136–65; galactic, 157–65; and Johannes Kepler, vii–viii; Type Ia, 141–48, 163; Type Ib, 148–51, 163; Type II, 151–57, 163–64

Supernova Bethlehem, 148. *See also* Nova of 5 B.C.
Suslyga, Laurentius, 45
Synoptic Gospels, 6–7

"Tears of Saint Lawrence," 110–11
Tempel, Ernest, 130
Tempel-Tuttle, Comet, 130–32, 226
Temple of Jerusalem, destruction of, 8, 9
Tertullian, 168
Theta Aquilae, 246, 280, 284, 286
Thomas, 16
Titius Aristo, 21
Torah and Gospel of Matthew, 13, 14–15
Trexler, R. C., 171–72
Triple conjunctions, 199–218; of 989–979 B.C., 208–9; of 861–860 B.C., 209–10; of 821–820 B.C., 210–11, 214; of 563–562 B.C., 211, 215; of 523–522 B.C., 211–12, 215; of 146–145 B.C., 212, 215; of 7 B.C., 208, 212–13, 254–56
Tuttle, Horace, 130
Tycho Brahe, 160–61, 162–63, 233, 240–41

Uranus, 99–100, 201
Urreta, Luis de, 172, 174

V500 Aquilae, 273–74
Venera probes, 79, 80
Venus, 73–74, 75–82, 103, 294n.1; and Chinese records at time of Jesus' birth, 236–37; conjunction with Jupiter, 96–98, 189; physical conditions on, 77–79
Venus Tablet, 76, 183, 187–88
Vespasian, Emperor of Rome, 226
Vesta, 100

Walker, Christopher, 182, 183, 194, 197
Wells, observation of stars from, 36–37
White dwarf star, 145–47, 233
William of Normandy, 87–90

Xi Ze-zong, 273

Yahweh, 14
Yeomans, Donald, 227, 239
Yuletide, 58–59

Zealots, 9
Zenith, 35, 36
Zoroastrianism and the Magi, 176, 179, 192–95